Studies in Computational Intelligence

Volume 578

Series editor

Janusz Kacprzyk, Polish Academy of Sciences, Warsaw, Poland
e-mail: kacprzyk@ibspan.waw.pl

About this Series

The series "Studies in Computational Intelligence" (SCI) publishes new developments and advances in the various areas of computational intelligence—quickly and with a high quality. The intent is to cover the theory, applications, and design methods of computational intelligence, as embedded in the fields of engineering, computer science, physics and life sciences, as well as the methodologies behind them. The series contains monographs, lecture notes and edited volumes in computational intelligence spanning the areas of neural networks, connectionist systems, genetic algorithms, evolutionary computation, artificial intelligence, cellular automata, self-organizing systems, soft computing, fuzzy systems, and hybrid intelligent systems. Of particular value to both the contributors and the readership are the short publication time frame and the world-wide distribution, which enable both wide and rapid dissemination of research output.

More information about this series at http://www.springer.com/series/7092

Roger Lee

Editor

Software Engineering Research, Management and Applications

 Springer

Editor
Roger Lee
Software Engineering and Information
   Technology Institute
Central Michigan University
Mount Pleasant, MI
USA

ISSN 1860-949X   ISSN 1860-9503   (electronic)
ISBN 978-3-319-38402-3   ISBN 978-3-319-11265-7   (eBook)
DOI 10.1007/978-3-319-11265-7

Springer Cham Heidelberg New York Dordrecht London

Printed on acid-free paper

Springer is part of Springer Science+Business Media (www.springer.com)

Foreword

The purpose of the 12th International Conference on Software Engineering, Artificial Intelligence Research, Management and Applications (SERA 2014) held on August 31–September 4, 2014 in Kitakyushu, Japan, is aimed at bringing together scientists, engineers, computer users, and students to share their experiences and exchange new ideas and research results about all aspects (theory, applications, and tools) of Software Engineering Research, Management and Applications, and to discuss the practical challenges encountered along the way and the solutions adopted to solve them. The conference organizers selected the best 17 papers from those papers accepted for presentation at the conference in order to publish them in this volume. The papers were chosen based on review score submitted by the members of the program committee and underwent further rigorous rounds of review.

In "Performance Impact of New Interface for Nonvolatile Memory Storage", Shuichi Oikawa proposes constructing a file system directly on it, and investigates the performance impact of different NVM storage characteristics for the proposed architecture. The evaluation results show that the proposed architecture is advantageous with the performance that can be realistically achieved by future NVM storage.

In "Optimized Test Case Generation Based on Operational Profiles with Fault-Proneness Information", Tomohiko Takagi and Mutlu Beyazıt present a novel formal model called an OPFPI (operational profile with fault-proneness information) and a novel algorithm to generate optimized test cases from the OPFPI are proposed in order to effectively improve software reliability or gain a perspective of software reliability in a limited span of time for testing.

In "Design Pattern for Self-Adaptive RTE Systems Monitoring", Mouna Ben Said, Yessine Hadj Kacem, Mickaël Kerboeuf, Nader Ben Amor, and Mohamed Abid define patterns for the monitoring and analyzing processes through the generalization of relevant existing adaptation approaches to improve their accessibility to new adaptive systems developers.

In "Changing the Software Engineering Education: A Report from Current Situation in Mexico", García, I., Pacheco, C., and Calvo-Manzano, J. present a strategy and federal programs established to modernize the software engineering curriculum in Mexico.

In "Does a Code Review Tool Evolve as the Developer Intended?", Osamu Mizuno and Junwei Liang try to answer the question "Does Rietveld evolve into Gerrit as the developers intended?" The result of analysis showed us that the improvements of Gerrit that the developer is expected are not observed explicitly.

In "Analysis of Mouth Shape Deformation Rate for Generation of Japanese Utterance Images Automatically", Tsuyoshi Miyazaki, and Toyoshiro Nakashima present research on machine lip-reading. In the process of the study, they propose a reproduction method of utterance images without voice from Japanese kana.

In "Predicting Fault-Prone Modules by Word Occurrence in Identifiers", Naoki Kawashima and Osamu Mizuno compare the word occurrence model with traditional models using CK metrics and LOC. The result of comparison showed that the occurrence of words is a good prediction measure as well as CK metrics and LOC.

In "The Influence of Alias and References Escape on Java Program Analysis", Shengbo Chen, Dashen Sun, and Huaikou Miao propose a static approach to detecting control flow information of programs with alias and references escape. A case study shows that the proposed method can detect control flow information exactly.

In "Preliminary Evaluation of a Software Security Learning Environment", Atsuo Hazeyama and Masahito Saito propose a learning process for software security and a learning environment that supports the learning process. They conducted a preliminary experiment to evaluate the learning process and the learning environment.

In "Expression Strength for the Emotional Scene Detection from Lifelog Videos", Atsushi Morikuni, Hiroki Nomiya, and Teruhisa Hochin propose a criterion to measure the strength of emotions for the purpose of retrieving emotional scenes from a lifelog video database, and introduce a criterion called expression strength in order to measure the strength of emotions based on the amount of the change of several facial feature values.

In "A Procedure for the Development of Mobile Applications Software", Byeongdo Kang, Jongseok Lee, Jonathan Kissinger, and Roger Y. Lee present a development method for mobile applications. Their method consists of five development phases. The results were quite promising and warrant further research.

In "The Relationship Between Conversation Skill and Feeling of Load on Youth to Communicate with Elderly Persons Using Video Image and Photographs", Miyuki Iwamoto, Noriaki Kuwahara, Kazunari Morimoto present the relationship between conversation skill and feeling of load on youth to communicate with elderly persons using video image and photographs.

In "Automatic Matching on Fracture Surface of Quarried Stone Using the Features of Plug-and-Feather Holes", Morita Karin, Ikeda Sei, and Sato Kosuke propose a novel method for matching of quarried blocks of a stone based on feature extraction of plug-and-feather holes. The experiments for an actual stone show the advantage of feature detection in automatic matching of quarried stones.

In "A Route Recommender System Based on the User's Visit Duration at Sightseeing Locations", Daisuke Kitayama, Keisuke Ozu, Shinsuke Nakajima, Kazutoshi Sumiya propose a route recommender system whose recommendations are based on the difference between a user's and the typical visit duration at sightseeing locations. In this study, they developed a prototype system and evaluated it using a few sightseeing locations.

In "Predicting Access to Health Care Using Data Mining Techniques", Sergey Shishlenin and Gongzhu Hu propose a method for predicting access to health care using data mining. They examine individual demand of receiving health care using the data from the Behavioral Risk Factor Surveillance System, a sample of the US population.

In "Object Tracking Method Using PTAMM and Estimated Foreground Regions", So, Hayakawa, Shinji Fukui, Yuji Iwahori, M. K. Bhuyan, and Robert J. Woodham propose a new approach for tracking moving objects in videos taken by a handheld camera. The effectiveness of the proposed method is shown through the experiments using real videos.

In "Application of Detecting Blinks for Communication Support Tool", Ippei Torii, Kaoruko Ohtani, Shunki Takami, Naohiro Ishii describe a new application operated with blinks for physically handicapped children who cannot speak to communicate with others. They develop the method into a communication application that has the accurate and the high-precision blink determination system to detect letters and put them into sound.

It is our sincere hope that this volume provides stimulation and inspiration, and that it will be used as a foundation for works to come.

July 2014   Osamu Mizuno
                                               Kyoto Institute of Technology, Japan

                                                                              Satoshi Takahashi
                                           University of Electro-Communications, Japan

                                                                                   Takaaki Goto
                                           University of Electro-Communications, Japan

Contents

Contributors

Mohamed Abid ENIS, CES Laboratory, University of Sfax, Sfax, Tunisia

Nader Ben Amor ENIS, CES Laboratory, University of Sfax, Sfax, Tunisia

Mouna Ben Said ENIS, CES Laboratory, University of Sfax, Sfax, Tunisia

Mutlu Beyazıt Faculty of Computer Science, Electrical Engineering and Mathematics, University of Paderborn, Paderborn, Germany

M.K. Bhuyan Indian Institute of Technology, Guwahati, India

J. Calvo-Manzano Computer Science School, Universidad Politecnica de Madrid, Madrid, Spain

Shengbo Chen School of Computer Engineering and Science, Shanghai University, Shanghai, China; Shanghai Key Laboratory of Computer Software Testing and Evaluating, Shanghai, China

Shinji Fukui Aichi University of Education, Kariya, Aichi, Japan

I. García Postgraduate Division, Universidad Tecnologica de la Mixteca, Huajuapan de León, Oaxaca, Mexico

Yessine Hadj Kacem ENIS, CES Laboratory, University of Sfax, Sfax, Tunisia

So Hayakawa Chubu University, Kasugai, Aichi, Japan

Atsuo Hazeyama Department of Information Science, Tokyo Gakugei University, Koganei-shi, Tokyo, Japan

Teruhisa Hochin Department of Information Science, Kyoto Institute of Technology, Sakyo-ku, Kyoto, Japan

Gongzhu Hu Department of Computer Science, Central Michigan University, Mount Pleasant, MI, USA

Sei Ikeda Osaka University, Toyonaka, Osaka, Japan

Naohiro Ishii Department of Information Science, Aichi Institute of Technology, Aichi, Japan

Yuji Iwahori Chubu University, Kasugai, Aichi, Japan

Miyuki Iwamoto Graduate School of Engineering and Science, Kyoto Institute of Technology, Kyoto, Japan

Byeongdo Kang Department of Computer and Information Technology, Daegu University, Daegu, Republic of Korea

Naoki Kawashima Software Engineering Laboratory, Graduate School of Science and Technology, Kyoto Institute of Technology, Kyoto, Japan

Mickaël Kerboeuf Lab-STICC, MOCS Team, University of Brest, Brest, France

Jonathan Kissinger Department of Computer Science, Central Michigan University, Mount Pleasant, MI, USA

Daisuke Kitayama Faculty of Information Science and Engineering, Kogakuin University, Shinjuku, Tokyo, Japan

Noriaki Kuwahara Graduate School of Engineering and Science, Kyoto Institute of Technology, Kyoto, Japan

Jongseok Lee Department of Computer Engineering, Woosuk University, Wanju-gun, Republic of Korea

Roger Y. Lee Department of Computer Science, Central Michigan University, Mount Pleasant, MI, USA

Junwei Liang Software Engineering Laboratory, Graduate School of Science and Technology, Kyoto Institute of Technology, Kyoto, Japan

Huaikou Miao School of Computer Engineering and Science, Shanghai University, Shanghai, China

Tsuyoshi Miyazaki Department of Information and Computer Sciences, Kanagawa Institute of Technology, Kanagawa, Atsugi, Japan

Osamu Mizuno Software Engineering Laboratory, Graduate School of Science and Technology, Kyoto Institute of Technology, Kyoto, Japan

Atsushi Morikuni Department of Information Science, Kyoto Institute of Technology, Sakyo-ku, Kyoto, Japan

Kazunari Morimoto Graduate School of Engineering and Science, Kyoto Institute of Technology, Kyoto, Japan

Karin Morita Osaka University, Toyonaka, Osaka, Japan

Shinsuke Nakajima Faculty of Computer Science and Engineering, Kyoto Sangyo University, Kyoto, Japan

Toyoshiro Nakashima School of Culture-Information Studies, Sugiyama Jogakuen University, Nagoya, Aichi, Japan

Hiroki Nomiya Department of Information Science, Kyoto Institute of Technology, Sakyo-ku, Kyoto, Japan

Kaoruko Ohtani Department of Information Science, Aichi Institute of Technology, Aichi, Japan

Shuichi Oikawa Faculty of Engineering, Information and Systems, University of Tsukuba, Tsukuba, Ibaraki, Japan

Keisuke Ozu Meitec Corporation, Fujisawa, Kanagawa, Japan

C. Pacheco Postgraduate Division, Universidad Tecnologica de la Mixteca, Huajuapan de León, Oaxaca, Mexico

Masahito Saito Graduate School of Education, Tokyo Gakugei University, Koganei-shi, Tokyo, Japan

Kosuke Sato Osaka University, Toyonaka, Osaka, Japan

Sergey Shishlenin Department of Computer Science, Central Michigan University, Mount Pleasant, MI, USA

Kazutoshi Sumiya Faculty of Human Science and Environment, University of Hyogo, Kobe, Hyogo, Japan

Dashen Sun School of Computer Engineering and Science, Shanghai University, Shanghai, China

Tomohiko Takagi Faculty of Engineering, Kagawa University, Takamatsu-shi, Kagawa, Japan

Shunki Takami Department of Information Science, Aichi Institute of Technology, Aichi, Japan

Ippei Torii Department of Information Science, Aichi Institute of Technology, Aichi, Japan

Robert J. Woodham University of British Columbia, Vancouver, BC, Canada

Performance Impact of New Interface for Non-volatile Memory Storage

Shuichi Oikawa

Abstract Non-volatile memory (NVM) storage is becoming a main stream of storage devices for various segments from the high end to the low end markets. While its high performance fits the high end markets of enterprise storage, its lower power consumption fits the low end markets of mobile devices. As NVM storage becomes more popular, its form evolves from the one, which is compatible with HDDs, into those, which suit the market requirements more appropriately. Its interface, which connects NVM storage with systems, also evolves in order to improve the performance. There is a claim that the further improvement of NVM storage performance makes it better to poll a storage device to sense completion of access requests rather than to use interrupts. Polling based storage can have the same interface as memory since it processes access requests synchronously and such synchronous processing causes no interrupt that is necessary for asynchronous processing. This paper premises that NVM storage will be in a form of main memory, proposes constructing a file system directly on it, and investigates the performance impact of different NVM storage characteristics for the proposed architecture. The evaluation results show that the proposed architecture is advantageous with the performance that can be realistically achieved by future NVM storage.

Keywords Operating systems · Non-volatile memory · File systems

1 Introduction

Non-volatile memory (NVM) storage is becoming a main stream of storage devices for various segments from the high end to the low end markets. Flash memory is currently the most popular NVM, and the flash memory based storage device is called

---

In this paper, flash memory stands for NAND flash memory.

S. Oikawa (✉)
Faculty of Engineering, Information and Systems, University of Tsukuba,
1-1-1 Tennodai, Tsukuba, Ibaraki, Japan
e-mail: shui@cs.tsukuba.ac.jp

© Springer International Publishing Switzerland 2015   1
R. Lee (ed.), *Software Engineering Research, Management and Applications*,
Studies in Computational Intelligence 578, DOI 10.1007/978-3-319-11265-7_1

an SSD (Solid State Drive). While its high performance fits the high end markets of enterprise storage, its lower power consumption fits the low end markets of mobile devices. As NVM storage becomes more popular, its form evolves from the one, which is compatible with HDDs, into those, which suit the market requirements more appropriately. Its interface, which connects NVM storage with systems, also evolves in order to improve the performance. The examples of the currently employed interfaces are PCI Express and NVM Express.

The investigation results shown by [1, 9] posed one interesting claim that the further improvement of NVM storage performance makes it better to poll a storage device to sense completion of access requests rather than to use interrupts. They investigated high performance NVM storage architecture and found that the existing block device interface of the operating system (OS) kernel does not always perform well with it. The reason of this claim is that by excluding the time required for process context switching and interrupt processing there is no time left for a yielded process to be executed if processing times of access requests shorten; thus, it is simply faster for systems to poll a device in order to wait for completion of access requests.

Such polling based storage can have the same interface as memory since it processes access requests synchronously and such synchronous processing causes no interrupt that is necessary for asynchronous processing; thus, NVM storage can evolve to be in a form of main memory in this aspect. Flash memory is, however, not byte addressable; thus, byte addressable RAM buffer cache is placed on flash memory in order to connect flash memory to the memory bus of CPUs, and address translation is performed to export the whole address space of flash memory. In this way, such storage devices becomes byte addressable, and CPUs can simply access the data on them by using memory access instructions. This paper calls this memory architecture *NV main memory* (*NVMM*) *storage*. Using NVMM storage as main memory, however, has a significant drawback since there is no way to predict addresses of future access; thus, access locality is the only means to mitigate the long access latency of NVM, and accessing uncached data can lead to significant performance degration.

This paper proposes constructing a file system directly on NVMM storage, so that it can reduce the software overhead to access storage by making use of its simple memory interface. The paper investigates the performance impact of different NVMM storage characteristics for the proposed architecture to find out the ideal specification of NVMM storage. The evaluation results show that the RAM buffer cache of NVMM storage can mitigate the access latency of NVM and also that the proposed architecture can take full advantage of the performance that can be realistically achieved by future NVMM storage.

The rest of this paper is organized as follows. Section 2 describes the background of the work. Section 3 describes the details of the proposed architecture. Section 4 describes the result of the experiments. Section 5 discusses the specification of the proposed architecture. Section 6 describes the related work. Section 7 summarizes the paper.

2 Background

This section describes the background of the work. It first describes the OS kernel's interaction with block devices. It then describes, the implication of NVM storage performance improvement. It finally describes the overall architecture of NVMM storage.

2.1 Interacting with Block Devices

Block devices, such as SSDs and HDDs, are not byte addressable; thus, CPUs cannot directly access the data on these devices. A certain size of data, which is typically multiples of 512 byte, needs to be transferred between memory and a block device for CPUs to access the data on the device. Such a unit to transfer data is called a block.

The OS kernel employs a file system to store data in a block device. A file system is constructed on a block device, and files are stored in it. In order to read the data in a file, the data first needs to be read from a block device to memory. If the data on memory was modified, it is written back to a block device. A memory region used to store the data of a block device is called a page cache. Therefore, CPUs access a page cache on behalf of a block device. Figure 1 depicts the hierarchy of CPUs, a page cache, and a block device as the existing architecture to interact with block devices.

Since HDDs are orders of magnitude slower than memory to access data on them, various techniques were devised to amortize the slow access time. The asynchronous access command processing is one of commonly used techniques. Its basic idea is that a CPU executes another process while a device processes a command. In Fig. 2, Process 1 issues a system call to access data on a block device. The kernel processes the system call and issues an access command to the corresponding device. The kernel then looks for the next process to execute and perform context switching to Process 2. Meanwhile, the device processes the command, and sends an interrupt to notify its completion. The kernel handles the interrupt, processes command completion, and performs context switching back to Process 1. T_{proc2} is a time left for Process 2 to

Fig. 1 The existing architecture to interact with block devices

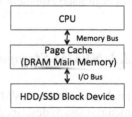

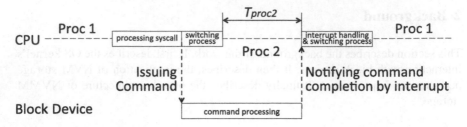

Fig. 2 The asynchronous access command processing and process context switches

run. Because HDDs are slow and thus their command processing time is long, T_{proc2} is long enough for Process 2 to proceed its execution.

2.2 Implication of NVM Storage Performance Improvement

The performance of flash memory storage has been improved by exploiting parallel access to multiple chips [3], and newer NVM technologies inherently achieve higher performance. Such higher performance changes a premise that devised the asynchronous access command processing to amortize the slow access time of block devices, and can affects how the OS kernel manages the interaction with NVM storage.

One of such possibilities is a claim made by [1, 9]. The claim compares the asynchronous access command processing with the synchronous processing, and shows that the synchronous processing is faster. The claim was supported by the experiments performed by employing a DRAM-based prototype block storage device, which was connected to a system through the PCIe Gen2 I/O bus. For a 4 KB transfer experiment performed by [9], the asynchronous processing takes 9.0 μs for Process 1 to receive data from the block device, and T_{proc2} is 2.7 μs. In contrast, the synchronous processing takes only 4.38 μs. The difference is 4.62 μs, which is longer than T_{proc2} of the asynchronous processing; thus, the synchronous processing saves the overall processing time.

As the I/O bus becomes faster, the command processing time of a device becomes shorter; thus, T_{proc2} also becomes shorter because the context switching time remains the same. Therefore, there will be no useful time left for another process while waiting for command completion.

2.3 NVMM Storage

High performance NVM storage makes asynchronous processing of access requests meaningless as described above. Moreover, most of the current NVM storage devices equip with a certain amount of RAM as buffer cache to accommodate transient data.

Fig. 3 The architecture
of NVMM storage

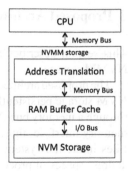

Though the sizes of RAM buffer cache vary in accord with their targets and prices, devices with 1 GB of RAM buffer cache can be found among recent products.

Such facts easily make NVM storage expand to become main memory. By making it directly connect to CPUs through a memory bus, there is no complex mechanism required to enable the asynchronous access command processing, and access requests are simply processed synchronously. In order to make NVM storage work as main memory, the whole address space made available by NVM storage needs to be addressable by CPUs while NVM storage cannot be directly connected to CPUs. Since the RAM buffer cache is connected to CPUs through an address translation mechanism, the address translation mechanism enables mapping of RAM to NVM storage. This paper calls this memory architecture *NVMM storage*, and Fig. 3 depicts it. Making NVM storage work as main memory is not a new idea as it was researched when flash memory appeared. eNVy [7] is such an example.

Newer NVM technologies, such as PCM and ReRAM, are byte addressable and provide higher performance than flash memory. They, however, have limited write similar to flash memory; thus, they cannot simply replace RAM, and the RAM buffer cache is required to place frequently accessed data. Therefore, NVMM storage is also legitimate for such newer NVM technologies.

This approach, however, has a significant drawback that access latencies vary depending upon the locations of the accessed data. If the data is on the RAM buffer cache, the access latencies are comparable to RAM. If the data is on NVM, the access latencies can be significantly longer. The problem is that there is no generally working way to predict addresses of future access. If addresses of future access can be predicted, the data of these addresses can be read ahead in the RAM buffer cache in order to mitigate the long access latency of NVM. Main memory is accessed by physical addresses. Because physical addresses do not provide any information of higher abstractions that can be clue to predict addresses of future access, no useful information is available. Therefore, access locality is the only means to accommodate frequently accessed data in the RAM buffer cache.

3 Proposed Architecture

This section describes the proposed architecture that constructs a file system on NVMM storage.

The proposed architecture utilizes NVMM storage as a base device of a file system. NVM storage persistently stores the data of a file system, and CPUs access the data through a memory bus. The file system directly interacts with the device through the memory interface; thus, the block device driver for NVM storage is not required. Files in the file system can be directly accessed and also mapped into the virtual address spaces of processes since files reside on memory; thus, there is no need to copy the data of files to a page cache. Therefore, the page cache mechanism is not required, either.

This architecture can be a solution to avoid the drawback posed by NVMM storage. File systems are designed to allocate the blocks referenced by a single file as contiguous as possible since contiguous blocks can be accessed faster on HDDs. Such contiguous blocks make it easy to read ahead blocks in the RAM buffer cache. Moreover, recent advances of file systems and storage architecture bring the concept of object based storage devices (OSDs), and higher abstractions are introduced and made available in storage. Such an architecture also enables to take advantage of the knowledge of higher abstractions for the prediction of future access. It should also be possible for user processes to give hints to NVMM storage in order to read ahead blocks since user processes should have knowledge of their access patterns.

The architecture also has several advantages over the existing file system architecture based on block devices, such as the simplification of the kernel architecture and faster processing of data access requests because of the simplified execution paths in the kernel. The kernel architecture can be significantly simplified since the architecture does not require block device drivers and a page cache. File systems directly interact with NVM storage through the memory interface although additional command interface may be needed to communicate with it in order to exchange hint information. The simplified architecture can stimulate active development of more advance features that take advantage of NVMM storage. Accessing data in a file becomes much faster since there is no need to go through a complex software framework that consists of a page cache mechanism and block device drivers in the kernel. Such a complex software framework paid off when block devices are as slow as HDDs. High performance NVM storage makes it outdated, and can rather take advantage of the simplified execution paths in the kernel.

4 Experiments

This section describes the experiment results that investigate the performance impact of different NVMM storage characteristics for the proposed architecture. We performed two experiments in order to see the performance impacts caused by the access latencies and concurrent request processing of NVMM storage.

4.1 Experiment Methodology

Since there is no physical NVMM storage device that has been developed so far, we developed a RAM disk device driver that emulate NVMM storage in order to perform experiments. Since experiments requires the emulation of NVMM storage with different characteristics, the developed driver takes the parameters for these characteristics. The parameters that can be specified include the access synchronization mode, the cache mode, the access latencies, and the number of requests that can be concurrently processed.

All experiments described below were performed on a PC-AT compatible system, which is equipped with an Intel Core i7-920 2.66GHz CPU. Execution times were measured by using the RDTSC instruction. The developed driver is installed in the Linux kernel, of which version is 3.4. The file system used for all experiments is the ext2 file system since it is the only XIP-enabled file system, which supports both read and write, in the main line Linux kernel 3.4.

For the measurements of file reading, the file was read when its data is not cached in the page cache managed by the kernel; thus, the measurements include the costs of page cache allocation if applicable. For the measurements of file writing, the file was written when no block of the file was allocated; thus, the measurements include the block allocation costs of a file system.

4.2 Performance Impact of Access Latency

The first experiment investigates the performance impact of the access latencies of NVMM storage. We performed the measurements of file reading and writing costs with the different access latencies emulated by the RAM disk device driver. The access synchronization modes used for the measurements are (1) the synchronous and direct access with XIP enabled without I/O request queueing (direct-xip), which is the proposed architecture, (2) the synchronous access with I/O request queueing (queueing-sync), and (3) the asynchronous access with I/O request queueing (queueing-async), which is the original access method of the block device drivers. There is the RAM buffer cache of NVMM storage, and the cache mode of the driver specifies if it is enabled or disabled. While file reading costs were measured with cache disabled, file writing costs were measured with cache enabled and disabled. For the queueing-async mode with cache disabled, the number of requests that can be concurrently processed was set to one (req = 1). The access latencies used for the measurements were from 0 to 16 μs. The queueing-async mode was not measured when there is no access latency. Because data is immediately available when there is no access latency, the asynchronous access mode is meaningless.

Figure 4 shows the measurement results of reading a 512MB file. From 0 to 16 μs, there is a constant difference between the direct-xip mode and the queueing-sync mode. For 2 μs of the access latency, the queueing-sync mode outperforms

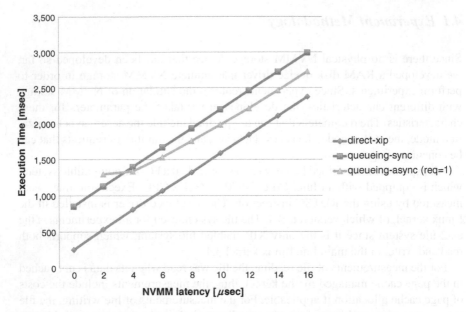

Fig. 4 The experiment result of reading a 512 MB file with various NVMM latencies

the queueing-async mode. For 4 μs and more of the access latency, however, the queueing-async mode outperforms the queueing-sync mode. As the access latency becomes larger, the difference of the execution time between them becomes larger, and the execution time of the queueing-async mode becomes closer to the direct-xip mode. For 16 μs of the access latency, there is still a certain difference of the execution time between the direct-xip mode and the queueing-async mode.

Figure 5 shows the measurement results of writing a 512 MB file. It is obvious that the execution times of NVMM storage with cache enabled are mostly constant and not affected by the access latencies, and that the direct-xip mode outperforms the other modes. For NVMM storage with cache disabled, the direct-xip mode outperforms the other modes, too. The differences between the direct-xip mode and the queueing-sync mode are mostly constant from 2 to 16 μs, and the difference is larger for 0 μs. Except for 0 μs, the characteristics of the direct-xip mode and the queueing-sync mode for writing is the same as reading. Only the queueing-async mode with cache disabled behaves differently for writing from reading since the increase of the access latencies of NVMM storage less affects the execution times. This characteristics can be because of the page cache of the OS kernel since the evictions from the page cache create the corresponding write requests and such evictions occur asynchronously from writing activities to a file. The combination of the two asynchronous mechanisms may cause synergy to absorb the effect of the increase of the access latencies.

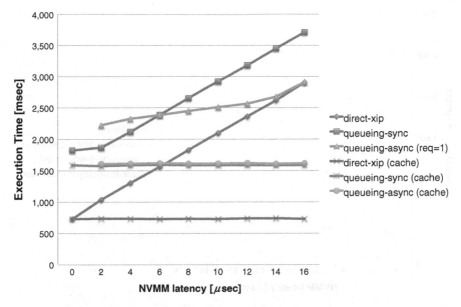

Fig. 5 The experiment result of writing a 512 MB file with various NVMM latencies

4.3 Performance Impact of Concurrent Request Processing

While the previous experiment set the number of requests that can be concurrently processed for the queueing-async mode, the number of such requests can be larger for real storage devices. For example, the native command queueing (NCQ) of the serial ATA (SATA) protocol allows the device driver to put multiple requests in the command queue of an SATA device; thus, the device can simultaneously process the requests if possible. Therefore, the second experiment investigates the performance impact of such concurrent request processing of NVMM storage, and we performed the measurements of file reading and writing costs with the different number of requests that can be concurrently processed for the queueing-async mode. The concurrent request processing is also emulated by the RAM disk device driver.

Figures 6 and 7 show the measurement results of reading and writing a 512 MB file, respectively. The results shows that the concurrent request processing reduces the execution times of the queueing-async mode significantly. For read, the queueing-async mode performs comparably with the direct-xip mode from 10 μs of the access latency when the number of concurrent requests is set to the maximum allowed by the block device driver framework. For write, the queueing-async mode outperforms the direct-xip mode from 8 μs of the access latency when the number of concurrent requests is set to 8.

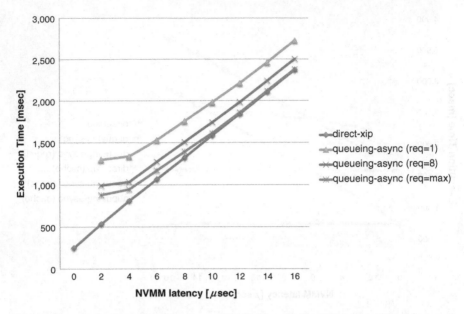

Fig. 6 The experiment result of reading a 512 MB file with the different numbers of requests processed concurrently

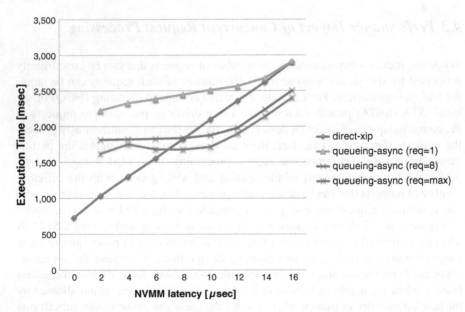

Fig. 7 The experiment result of writing a 512 MB file with the different numbers of requests processed concurrently

5 Discussion

This section discusses the specification of NVMM storage for the direct-xip mode to outperform the other modes that employ the I/O request queueing.

For read, it is better for NVMM storage to employ RAM buffer cache if the access latency is more than $10\,\mu s$. The RAM buffer cache, however, needs to be intelligently controlled to read ahead data that will be accessed soon. Otherwise, the cache does not contribute to the performance.

For write, NVMM storage requires the RAM buffer cache if the access latency is more than $6\,\mu s$. The RAM buffer cache does not need to be controlled very intelligently for write. Because it temporarily buffers written data, the data simply needs to be copied to the storage device.

In summary, NVMM storage requires the RAM buffer cache if the access latency of the storage device is more than $6\,\mu s$. The cache needs to be intelligently controlled for effective read ahead. If the access latency of the storage device is less than $6\,\mu s$, the RAM buffer cache is not necessary. While it is difficult to achieve the access latency of sub $10\,\mu s$ as long as flash memory is used as a storage device, the next generation NV memory performs much better than flash memory and should be achieve the access latency of sub $10\,\mu s$ [6].

6 Related Work

eNVy [7] proposes NV main memory system. It assumes the utilization of NOR flash memory, which is byte addressable and of which read access latency is comparable to DRAM; thus, the device characteristics allow its use as the main memory, and did not consider the construction of a file system on it. Moneta [1] is a storage array architecture designed for NVM. While its evaluation revealed the necessity of reducing the software costs to deal with block devices, it does not consider the removal of the block device interface. Yang et. al. [9] also investigated the software costs to deal with block devices on the premise that PCIe gen3 based flash storage devices will become even faster. While it proposed the synchronous interface to block devices, it is still based on the block device interface. Tanakamaru et. al. [6] proposed a storage state storage device that is a hybrid of flash memory and ReRAM. Their hybrid architecture is similar to the NVMM storage, they do not discuss the interaction with a file system. Meza et. al. [5] describes the idea to coordinate the management of memory and storage under a single hardware unit in a single address space. They focused energy efficiency, and did not propose any software architecture. Condit et. al. [2] and SCMFS [8] proposed the file systems that were designed to be constructed directly on NVM. They, however, have no consideration of hybrid storage architecture.

The memory channel storage (MCS) of Diabro Technologies [4], from which SanDisk productized ULLtraDIMM[1] as of writing, is a NAND flash memory based storage device that can be connected to the memory bus of processors. MCS is, however, not byte-addressable main memory but a block device. In conjunction with the assist provided by the firmware to emulate a block device, the existing block device driver can be used to access MCS with no modification. If MCS is modified to be byte-addressable main memory, our proposed software architecture will certainly be able to make use of it.

7 Summary and Future Work

Non-volatile memory (NVM) storage is becoming a main stream of storage devices for various segments from the high end to the low end markets. High performance NVM storage can can have the same interface as memory since it processes access requests synchronously and such synchronous processing causes no interrupt that is necessary for asynchronous processing; thus, NVM storage can evolve to be in a form of main memory. This paper premises that NVM storage will be in a form of main memory, proposes constructing a file system directly on it, and investigated the performance impact of different NVM storage characteristics for the proposed architecture. The evaluation results showed that the proposed architecture is advantageous with the performance that can be realistically achieved by future NVM storage.

We are currently developing an intelligent caching method in order to read ahead data that will be accessed soon by cooperating with a file system.

References

1. Caulfield, A.M., De, A., Coburn, J., Mollow, T.I., Gupta, R.K., Swanson, S.: Moneta: A high-performance storage array architecture for next-generation, non-volatile memories. In: Proceedings of the 2010 43rd Annual IEEE/ACM International Symposium on Microarchitecture, MICRO '43, pp. 385–395. IEEE Computer Society, Washington, DC, USA (2010). doi:10.1109/MICRO.2010.33. http://dx.doi.org/10.1109/MICRO.2010.33
2. Condit, J., Nightingale, E.B., Frost, C., Ipek, E., Lee, B., Burger, D., Coetzee, D.: Better i/o through byte-addressable, persistent memory. In: Proceedings of the ACM SIGOPS 22nd Symposium on Operating Systems Principles, SOSP '09, pp. 133–146. ACM, New York, NY, USA (2009). doi:10.1145/1629575.1629589. http://doi.acm.org/10.1145/1629575.1629589
3. Josephson, W.K., Bongo, L.A., Li, K., Flynn, D.: Dfs: a file system for virtualized flash storage. Trans. Storage 60(3), 14:1–14:25 (2010). doi:10.1145/1837915.1837922. http://doi.acm.org/10.1145/1837915.1837922
4. Memory channel storage. http://www.diablo-technologies.com/ (2013)
5. Meza, J., Luo, Y., Khan, S., Zhao, J., Xie, Y., Mutlu, O.: A case for efficient hardware-software cooperative management of storage and memory. In: Proceedings of the 5th Workshop on Energy-Efficient Design (WEED), pp. 1–7 (2013)

---

[1] ULLtraDIMM is used by IBM under the eXFlash DIMM brand name.

6. Tanakamaru, S., Doi, M., Takeuchi, K.: Unified solid-state-storage architecture with nand flash memory and reram that tolerates 32x higher ber for big-data applications. In: 2013 IEEE International Solid-State Circuits Conference Digest of Technical Papers (ISSCC), pp. 226–227 (2013). doi:10.1109/ISSCC.2013.6487711
7. Wu, M., Zwaenepoel, W.: eNVY: a non-volatile, main memory storage system. In: Proceedings of the 6th International Conference on Architectural Support for Programming Languages and Operating Systems, ASPLOS VI, pp. 86–97. ACM, New York, NY, USA (1994). doi:10.1145/195473.195506. http://doi.acm.org/10.1145/195473.195506
8. Wu, X., Reddy, A.L.N.: Scmfs: a file system for storage class memory. In: Proceedings of 2011 International Conference for High Performance Computing, Networking, Storage and Analysis, SC '11, pp. 39:1–39:11. ACM, New York, NY, USA (2011). doi:10.1145/2063384.2063436. http://doi.acm.org/10.1145/2063384.2063436
9. Yang, J., Minturn, D.B., Hady, F.: When poll is better than interrupt. In: Proceedings of the 10th USENIX Conference on File and Storage Technologies, FAST'12, pp. 1–7. USENIX Association, Berkeley, CA, USA (2012). http://dl.acm.org/citation.cfm?id=2208461.2208464

Optimized Test Case Generation Based on Operational Profiles with Fault-Proneness Information

Tomohiko Takagi and Mutlu Beyazıt

Abstract In this paper, a novel formal model called an OPFPI (operational profile with fault-proneness information) and a novel algorithm to generate optimized test cases from the OPFPI are proposed in order to effectively improve software reliability or gain a perspective of software reliability in a limited span of time for testing. The OPFPI includes the feasibility problem due to the use of guards (conditions to make specific state transitions feasible); therefore, ant colony optimization is employed in the algorithm to generate test cases that cover frequent and fault-prone state transitions as comprehensively as possible. A test tool that implements the OPFPI and executes the optimized test case generation has been developed, and it has been applied to a non-trivial system. The obtained results indicate that significant improvement of test cases can be achieved in a short time.

Keywords Model-based software testing · Operational profile · Test case generation · Ant colony optimization

1 Introduction

Testing is an important process to detect faults in software before shipping, and it can be performed systematically by using different software testing techniques. The testing techniques, however, cannot assure that there are no faults in the software [1], but software reliability [11] can be improved by correcting the detected faults. Software reliability is concerned with the operation of the software in expected

T. Takagi (✉)
Faculty of Engineering, Kagawa University, 2217-20 Hayashi-cho,
Takamatsu-shi, Kagawa 761-0396, Japan
e-mail: takagi@eng.kagawa-u.ac.jp

M. Beyazıt
Faculty of Engineering, Department of Computer Engineering,
Yaşar University, Üniversite Caddesi, No: 37–39, Ağaçli Yol, Bornova, Izmir, Turkey
e-mail: mutlu.beyazit@yasar.edu.tr

© Springer International Publishing Switzerland 2015   15
R. Lee (ed.), *Software Engineering Research, Management and Applications*,
Studies in Computational Intelligence 578, DOI 10.1007/978-3-319-11265-7_2

environments without any faults; it is determined by *usage distributions* (i.e., how software is used by users or other systems) and *fault distributions* (i.e., how software includes latent faults).

OPBT (*operational profile-based testing*) [9, 14] is a testing technique focusing on usage distributions; it is sort of a combination of MBT (model-based testing), random testing, and state-transition-based testing. The operational profile is a probabilistic state machine created by a test engineer; it consists of a state machine and probability distributions. The former represents the expected behavior of SUT (software under test), and it is constructed based on software specifications. The latter represents the characteristics of the expected usage of the SUT, and it is derived based on the survey of operational environments. The operational profile is used to randomly generate test cases in the form of state-transition sequences starting from an initial state. The test cases statistically reflect the characteristics of the expected usage in operational environments, and, therefore, OPBT is effective in detecting faults that relate to frequent usage in operational environments and have serious impacts on software reliability. Thus, OPBT is generally not suitable for detecting the faults that relate to infrequent usage.

FABT (*fault analysis-based testing*) [5, 13], on the other hand, focuses on fault distributions. In FABT, the complexity of software specification or source code is evaluated by software metrics, or bug reports stored in a bug tracking system are analyzed by mathematical techniques. Then, test cases are created to test fault-prone parts that have high complexity or relate to a large number of bug reports. FABT is effective in devoting the test effort to the parts that have the likelihood of including a number of latent faults, but it does not ensure that test cases detect faults that have serious impacts on software reliability.

Test engineers require a testing technique that has the strength of both OPBT and FABT in order to effectively improve the software reliability or gain an insight into it in a limited span of time for testing. To our knowledge, such a testing technique has not been established except for [3], which, however, employs more of an analytical approach and uses simulations to support the analysis; it does not employ any models or propose any test case generation algorithms.

In this paper, we propose a new formal model called an *OPFPI* (*operational profile with fault-proneness information*), and a novel algorithm to generate optimized test cases from the OPFPI. The generated test cases have to satisfy the various constraints of the state machine and the test tactics to cover both the transitions that are frequently executed in operational environments and the fault-prone transitions that have high complexity or relate to a large number of bug reports. To solve this constraint problem, we construct the algorithm using ACO (ant colony optimization) [12], which is a metaheuristic suitable for applying to test case generation in MBT [4, 8].

The rest of this paper is organized as follows. Section 2 discusses related work. In Sects. 3 and 4, an OPFPI model and an optimized test case generation algorithm based on OPFPI are proposed, respectively. Section 5 evaluates the proposed technique by discussing its effectiveness and limitations over a non-trivial system. Finally, Section 6 concludes the paper and outlines future work.

2 Related Work

OPBT was proposed as a software testing technique mainly for the rapid improvement of software reliability. There are several studies reporting its effectiveness. For example, Hartmann et al. reduce the test efforts for medical systems by introducing OPBT [7]. One of the challenges in OPBT is to derive the usage distributions, and they show that the usage distributions can be systematically derived from software execution histories captured by a test tool (a capture/replay tool). Chruscielski et al. construct operational profiles for software of aircrafts, and show that it is effective for improving the test process and the software reliability [2]. However, Beizer points out that software functions of frequent use have fewer faults and, therefore, test efforts should be concentrated on the ones of infrequent use rather than of frequent use [1]. In general, several factors can affect the fault detection ability of a software testing technique. Also, relationships between usage and fault distributions are not always straightforward, which is a reason why we focus on not only usage distributions but also fault distributions.

In OPBT, there is a formal model called a test model [14], which is similar to the OPFPI model proposed in this paper. The test model is a probabilistic state machine that provides an overview of test results and the way to evaluate test stopping criteria. The state machine is constructed based on the actual behavior of the SUT and the probability distributions derived from the results of test case execution. If a fault is detected by test case execution, a special state transition that represents the occurrence of the fault and its probability are added to the test model. When a sufficient number of test cases are executed and a certain number of faults are revealed, the difference between the probability distributions of the operational profile and the test model becomes small. This means that test engineers can stop the OPBT, because the software reliability of the SUT has reached a sufficient level. Unlike OPFPI, the test model of OPBT does not operate on pure usage distributions and detailed fault distributions. Therefore, it is relatively less suitable for test case generation.

It is a well-known fact that there are relationships between software metrics and software quality, and a practical test tactic is to concentrate test efforts on the parts that possess low quality from the perspective of software metrics. For example, Eski et al. propose a technique to predict parts of low quality by combining several object-oriented software metrics [5]. Whittaker et al. show the way to predict fault-prone parts by using the frequencies and time of bug fixes [13]. However, software metrics and fault analysis have not been incorporated into MBT that is effective in systematically testing large and complex software.

Most techniques of MBT are aimed at covering the model as comprehensive as possible, and some recent researches introduce metaheuristics in order to generate optimized test cases, that is, test cases that satisfy the constraints included in the model and exercise all elements of the model by using a smaller set of test inputs. For example, it is difficult to deterministically generate optimized test cases from a state machine that includes guards (conditions to make specific transitions feasible), which leads to a feasibility problem. GAs (genetic algorithms) can be used to solve the feasibility problem and generate optimized test cases that cover the transitions of the state machine [4, 8]. GAs are well-known metaheuristics that imitate the evolution of life to find approximate solutions. Srivastava et al. show that ACO works better than the GA on the feasibility problem [12]. ACO is a metaheuristic imitating the ants searching for food. Our technique is not aimed at covering the model, but it must address the feasibility problem since an OPFPI includes guards. Therefore, we employ ACO in our testing technique.

3 Operational Profile with Fault-Proneness Information

In this section, we propose an OPFPI, which is a novel formal model that consists of a conventional operational profile (a state machine and usage distributions) and fault distributions. An OPFPI is based on a state machine SM which is defined as follows.

Definition 1 $SM = \{S, E, s_0, F, \delta, G\}$, where:

- S is a finite set of states.
- E is a finite set of events corresponding to the stimuli from users and other systems, etc.
- s_0 is an initial state where $s_0 \in S$.
- F is a finite set of final states where $F \subseteq S$.
- δ is a state transition function where $\delta : S \times E \times \{G \cup \emptyset\} \rightarrow S$.
- G is a finite set of guards to express conditions that should be satisfied before specific state transitions can be performed.

An OPFPI is constructed by extending a state machine. The definition is given as follows:

Definition 2 $OPFPI = \{S, E, s_0, F, \delta, G, D_u, D_f\}$, where:

- D_u expresses usage distributions, that is, a finite set of occurrence probabilities of event e ($e \in E$) in state s ($s \in S$) in expected operational environments.
- D_f expresses fault distributions of the SUT, that is, a finite set of fault-proneness values of event e ($e \in E$) in state s ($s \in S$).

An example OPFPI is shown in Fig. 1. This OPFPI has three states and four events. In this paper, states and events are identified by numbers and letters, respectively. When the current state is state 1, events a, b, c and d can occur with probabilities of 25, 55, 10 and 10 %, and bring the current state to states 2, 1, 3 and 1, respectively. The fault-proneness values of these transitions, expressed as $(1, a)$, $(1, b)$, $(1, c)$ and $(1, d)$, are 10, 60, 5 and 10, respectively. Note that all the fault-proneness values in Fig. 1 are normalized to range from 0 to 100. $(1, b)$ has a high occurrence probability and a high fault-proneness value, and, therefore, test efforts should be intensively devoted to it in state 1. On the other hand, when the current state is state 2, the

event state	a event a	b event b	c event c	d event d
1	state 1 (initial state)	25% / 10 2	55% / 60 1	10% / 5 3	10% / 10 1
2	state 2	40% / 5 1[guard 1], 2[guard 2]	10% / 100 3	50% / 15 1[guard 2]	N/A
3	state 3	N/A	N/A	N/A	100% / 0 1

Fig. 1 Simple example of an OPFPI

evaluation of guards is required in order to determine feasible events. For example, when only guard 1 is satisfied in state 2, events a and b can occur with probabilities of 80 and 20%, and bring the current state to states 1 and 3, respectively. Note that event c cannot occur and becomes N/A, since guard 2 is not satisfied in this case. The fault-proneness values of $(2, a)$ and $(2, b)$ are 5 and 100, respectively. $(2, b)$ has a low occurrence probability but an extremely high fault-proneness value, and, therefore, it should be focused in testing of state 2.

An OPFPI can be constructed by the following steps.

(Step 1) Test engineers construct a state machine that represents the expected behavior of SUT using the software specifications. The abstraction level of the state machine is determined based on the purpose of testing.

(Step 2) Test engineers can derive usage distributions in the expected operational environments of the SUT by using the following investigation methods:

- The execution histories are collected from other software that is similar to the SUT (such as a prior version or prototype of the SUT) by test/debug/security tools, and the state machine constructed in Step 1 is traced using the collected information in order to calculate the occurrence probabilities of events in states [7].
- The questionnaire or the expectation of engineers and users determine the occurrence probabilities [2].
- The usage distributions of other operational profiles are reused by mathematical programming [10].

(Step 3) Test engineers can derive fault distributions by using software metrics and fault prediction techniques, such as LOC (lines of code), cyclomatic complexity, Halstead's software science, CK metrics, MOOD metrics, and the fault prediction based on the frequencies and time of bug fixes [1, 5, 13]. The software metrics are evaluated by existing software development tools, and, also, the fault prediction can be automatically performed based on bug reports stored in existing bug tracking systems. One or more software metrics and fault prediction techniques are selected based on the domain of the SUT and the purpose of testing to obtain fault-proneness values of events in states. When multiple software metrics and fault prediction techniques are selected, the results of them are normalized and combined [5]. For example, when metrics m_1 taking values in $[0, \infty)$ and m_2 taking values in $[0, 1]$ are selected, they can be normalized to take values in $[0, 100]$, and then combined by using the normalized values of m_1 and m_2.

4 Test Case Generation Using Ant Colony Optimization

In this section, we propose a novel algorithm for generating optimized test cases from a given OPFPI. ACO is introduced to effectively generate optimized test cases that satisfy the various constraints of the state machine and the following test tactics.

- Cover the transitions that will be frequently executed in operational environments, which are revealed by usage distributions in the OPFPI.
- Cover the fault-prone transitions that have high complexity or relate to a large number of bug reports, which are revealed by fault distributions in the OPFPI.

The basic idea of the algorithm is that agents (i.e., ants) perform reiteration search of paths (i.e., test cases) on the OPFPI based on usage distributions, fault distributions, and pheromones. The agents release and add the pheromones onto traversed paths in order to improve their future search. The details of the algorithm are as follows.

(Step 1) A ($A \geq 1$) agents whose movable distance (i.e., the number of transitions that a test case consists of) is D are created, and then the initial pheromone levels of all transitions of the OPFPI are set at τ_0 ($\tau_0 \geq 0.0$).

(Step 2)   All the agents perform the behavior specified by Step 2.1, 2.2, and 2.3.

(Step 2.1)   An agent is placed on the initial state of the OPFPI.

(Step 2.2)   The agent searches a path by repeating the random selection of a feasible event. The probability that an agent a traverse a transition by occurrence of an event e in a state s in time t (the number of execution of Step 2) is given as follows.

$$p_{se}^a(t) = \begin{cases} \dfrac{[\tau_{se}(t)] \cdot [\eta_{se}]^\alpha}{\sum_{x \in E_a(s)} [\tau_{sx}(t)] \cdot [\eta_{sx}]^\alpha} \,, & \text{if } e \in E_a(s) \\ 0 & , \quad \text{otherwise} \end{cases} \tag{1}$$

where $E_a(s)$ expresses a set of events that an agent a can select in a state s, that is, events that are not N/A and satisfy their guard conditions in a state s; $\tau_{se}(t)$ expresses a pheromone level of an event e in a state s in time t; η_{se} expresses a heuristic value of an event e in a state s; and α is a parameter to control the weighting between the pheromone level and the heuristic value. Also, η_{se} is calculated as follows.

$$\eta_{se} = \beta \cdot \frac{u_{se}}{\sum_{x \in E(s)} u_{sx}} + (1 - \beta) \cdot \frac{f_{se}}{\sum_{x \in E(s)} f_{sx}} \tag{2}$$

where $E(s)$ expresses a set of events that can occur in a state s, that is, events that are not N/A in a state s; u_{se} expresses occurrence probability of an event e in a state s in operational environments; f_{se} expresses a fault-proneness level of an event e in a state s; and β $(0.0 \leq \beta \leq 1.0)$ is a parameter to control the weighting between usage distributions and fault distributions.

(Step 2.3)   The agent completes searching a path when it reaches a final state of the OPFPI or the length of the searched path reaches the movable distance D.

(Step 3)   When $t = N$ $(N \geq 1)$, the best path among all the paths that have been searched by the agents is selected as a solution (an approximate optimal solution), and the algorithm terminates. Evaluation of a path P is performed as

$$E_P = \sum_{(s,e)\in P} \eta_{se} \tag{3}$$

where η_{se} expresses a heuristic value of an event e in a state s. Note that the same (s, e) can appear in P repeatedly, but its heuristic value is not added repeatedly. P with a larger value of E_P is better for achieving the test tactics mentioned before.

(Step 4)   For the next execution of Step 2, the pheromone levels of all the transitions are updated as follows.

$$\tau_{se}(t + 1) = (1 - \rho) \cdot \tau_{se}(t) + \sum_{a=1}^{A} \Delta\tau_{se}^{a} \tag{4}$$

where ρ $(0.0 < \rho < 1.0)$ expresses the evaporation rate of pheromones between t and $t + 1$. Furthermore, $\Delta\tau_{se}^{a}$ expresses pheromones that an agent a newly adds to (s, e), and it is calculated as

$$\Delta\tau_{se}^{a} = \begin{cases} E_{P_a}, & \text{if } (s, e) \in P_a \\ 0, & \text{otherwise} \end{cases} \tag{5}$$

where P_a expresses a path that have been searched by an agent a. After the update, the algorithm goes back to Step 2.

The parameters used in the above algorithm are A (the number of agents), D (the movable distance of agents), N (the number of execution of Step 2), τ_0 (the initial pheromone level), α (the parameter to control the weighting between the pheromone level and the heuristic value), β (the parameter to control the weighting between usage distributions and fault distributions), and ρ (the evaporation rate of pheromones). In practice, the values for these parameters are determined by trial-and-error. Thus, test engineers can generate test cases suitable for the domain of SUT and the purpose of testing by using this algorithm with properly selected parameter values.

5 Evaluation

In order to evaluate our technique, we developed a test tool that implements the OPFPI model and performs the optimized test case generation proposed in Sects. 3 and 4. A non-trivial system is used to show an overview of the application, to obtain results and to evaluate the proposed technique based on these results.

The system is software that controls a vending machine, and its requirements are shown in Fig. 2. We created an OPFPI for the software based on the requirements. Also, usage and fault distributions are assumed and given in Fig. 3. The constructed OPFPI has 16 states, 8 events, 82 transitions (i.e., the cells in Fig. 3 that do not contain N/A), and several guards. The model is non-trivial; it is sufficiently large to experiment and evaluate the testing method proposed in the paper. Note that Fig. 3 can also be expressed as a state transition diagram.

Optimized test cases are generated from the constructed OPFPI automatically by using the test tool with parameter values of $A = 1$, $D = 300$, $N = 500$, $\tau_0 = 0.1$, $\alpha = 1.0$, $\beta = 0.5$ and $\rho = 0.05$, which are determined by trial-and-error. Fig. 4 outlines a graph, where the horizontal axis shows t (time, i.e., the number of executions of Step 2) and the vertical axis shows E_P (the evaluation of a test case P—See Eq. (3)).

In Fig. 4, P in the thick line corresponds to the best test case among all the generated test cases; that is, the thick line demonstrates the growth of the best test case. On the other hand, P in the thin line corresponds to a test case that is newly generated for each t. The thick line reveals that E_P for $t = 1$ (i.e., in the case that the ACO is not executed) and for $t = 500$ (i.e., in the case that the ACO is executed) are about 7.3 points and 12.2 points respectively. It means that this optimized test case generation has achieved an improvement of about 4.9 points (about 68 %).

Our optimized test case generation is probabilistic; therefore, the generation with the above parameter settings was executed 100 times in order to obtain more reliable results. As a result, the average E_P of the best test case for $t = 500$ (i.e., final solution) is about 11.2 points, and its MAD (mean absolute deviation) is about 0.98. The MAD is small enough to conclude that the obtained results are stable; that is,

---

The VM (vending machine) sells coffees at 100 JPY (Japanese yen) apiece, and teas at 120 JPY apiece. It accepts only 100 JPY coins and 10 JPY coins. If the total of coins dropped into the VM reaches the price of a coffee or a tea, the button to buy a coffee or a tea becomes available respectively. If the available button is pushed, the VM provides a coffee or a tea, and the change (10 JPY coins). If a return lever is pressed down, the VM returns the dropped coins.

   The VM indicates the lack of coffees, teas, and 10 JPY coins for change. If coffees or teas are sold out, the corresponding button does not become available even if sufficient money to buy is dropped into the VM. Also, if sufficient money requiring the change is dropped into the VM and the VM does not have sufficient 10 JPY coins for the change, the button to buy a tea does not become available. When no money is dropped into the VM, coffees, teas, and 10 JPY coins can be supplied by a serviceman.

Fig. 2 Requirements of software to control a vending machine

		event a	b	c	d	e	f	g	h
state		drop a 10JPY coin	drop a 100JPY coin	return	buy a coffee	buy a tea	supply change	supply coffees	supply teas
1	b_coffee:off, b_tea:off / s_coffee:off, s_tea:off / change:off	25% / 20, 1 [drop<100JPY], 2 [drop>=100JPY]	55% / 20, 2 [drop<120JPY], 3 [drop>=120JPY]	5% / 0, 1 [drop>0JPY]	N/A	N/A	5% / 5, 1 [drop=0JPY]	5% / 5, 1 [drop=0JPY]	5% / 5, 1 [drop=0JPY]
2	b_coffee:on, b_tea:off / s_coffee:off, s_tea:off / change:off	30% / 20, 2 [drop<120JPY], 3 [drop=120JPY]	30% / 10, 3	5% / 5, 1	35% / 50, 1 [!g_coffee], 4 [g_coffee]	N/A	N/A	N/A	N/A
3	b_coffee:on, b_tea:off / s_coffee:on, s_tea:off / change:off	3% / 5, 3	2% / 5, 3	5% / 5, 1	20% / 55, 1 [!g_coffee], 4 [g_coffee]	70% / 85, 1 [!g_tea & !g_change], 6 [g_tea & !g_change], 9 [!g_tea & g_change], 14 [g_tea & g_change]	N/A	N/A	N/A
4	b_coffee:off, b_tea:off / s_coffee:on, s_tea:off / change:off	30% / 20, 4 [drop<120JPY], 5 [drop>=120JPY]	45% / 20, 4 [drop<120JPY], 5 [drop>=120JPY]	5% / 0, 4 [drop>0JPY]	N/A	N/A	5% / 5, 4 [drop=0JPY]	10% / 10, 1 [drop=0JPY]	5% / 5, 4 [drop=0JPY]
5	b_coffee:off, b_tea:on / s_coffee:on, s_tea:on / change:off	3% / 5, 5	2% / 5, 5	5% / 5, 4	N/A	90% / 85, 4 [!g_tea & !g_change], 8 [g_tea & !g_change], 12 [!g_tea & g_change], 16 [g_tea & g_change]	N/A	N/A	N/A
6	b_coffee:off, b_tea:off / s_coffee:off, s_tea:on / change:off	20% / 20, 6 [drop<100JPY], 7 [drop=100JPY]	55% / 15, 7 [drop>=100JPY]	5% / 0, 6 [drop>0JPY]	N/A	N/A	5% / 5, 6 [drop=0JPY]	5% / 5, 6 [drop=0JPY]	10% / 10, 1 [drop=0JPY]
7	b_coffee:on, b_tea:off / s_coffee:off, s_tea:on / change:off	3% / 5, 7	2% / 5, 7	5% / 5, 6	90% / 50, 6 [!g_coffee], 8 [g_coffee]	N/A	N/A	N/A	N/A
8	b_coffee:on, b_tea:on / s_coffee:on, s_tea:on / change:off	4% / 5, 8	6% / 5, 8	10% / 0, 8 [drop>0JPY]	N/A	N/A	10% / 5, 8 [drop=0JPY]	35% / 10, 6 [drop=0JPY]	35% / 10, 4 [drop=0JPY]
9	b_coffee:off, b_tea:off / s_coffee:off, s_tea:off / change:on	40% / 20, 9 [drop<100JPY], 10 [drop=100JPY]	20% / 30, 10 [drop<120JPY	10JPYcoins<2], 11 [drop>=120JPY & 10JPYcoins>=2]	10% / 0, 9 [drop>0JPY]	N/A	N/A	20% / 10, 1 [drop=0JPY]	5% / 5, 9 [drop=0JPY]	5% / 5, 9 [drop=0JPY]
10	b_coffee:on, b_tea:off / s_coffee:off, s_tea:off / change:on	30% / 30, 10 [drop<120JPY	10JPYcoins<2], 11 [drop>=120JPY & 10JPYcoins>=2]	5% / 20, 10 [10JPYcoins<2], 11 [10JPYcoins>=2]	10% / 5, 9	55% / 80, 1 [!g_coffee & !g_change], 4 [g_coffee & !g_change], 9 [!g_coffee & g_change], 12 [g_coffee & g_change]	N/A	N/A	N/A	N/A
11	b_coffee:off, b_tea:on / s_coffee:off, s_tea:off / change:on	3% / 5, 11	2% / 5, 11	10% / 5, 9	18% / 85, 1 [!g_coffee & !g_change], 4 [g_coffee & !g_change], 9 [!g_coffee & g_change], 12 [g_coffee & g_change]	67% / 100, 1 [!g_tea & !g_change], 6 [g_tea & !g_change], 9 [!g_tea & g_change], 14 [g_tea & g_change]	N/A	N/A	N/A
12	b_coffee:off, b_tea:off / s_coffee:on, s_tea:off / change:on	40% / 30, 12 [drop<120JPY	10JPYcoins<2], 13 [drop>=120JPY & 10JPYcoins>=2]	25% / 30, 12 [drop<120JPY	10JPYcoins<2], 13 [drop>=120JPY & 10JPYcoins>=2]	10% / 0, 12 [drop=0JPY]	N/A	N/A	10% / 10, 4 [drop=0JPY]	10% / 10, 9 [drop=0JPY]	5% / 5, 12 [drop=0JPY]
13	b_coffee:off, b_tea:on / s_coffee:on, s_tea:off / change:on	3% / 5, 13	2% / 5, 13	10% / 5, 12	N/A	85% / 85, 4 [!g_tea & !g_change], 8 [g_tea & !g_change], 12 [!g_tea & g_change], 16 [g_tea & g_change]	N/A	N/A	N/A
14	b_coffee:off, b_tea:off / s_coffee:off, s_tea:on / change:on	20% / 20, 14 [drop<100JPY], 15 [drop=100JPY]	45% / 10, 15	10% / 0, 14 [drop>0JPY]	N/A	N/A	10% / 10, 6 [drop=0JPY]	5% / 5, 14 [drop=0JPY]	10% / 10, 9 [drop=0JPY]
15	b_coffee:off, b_tea:on / s_coffee:off, s_tea:on / change:on	3% / 5, 15	2% / 5, 15	10% / 5, 14	N/A	85% / 80, 6 [!g_coffee & !g_change], 8 [g_coffee & !g_change], 15 [!g_coffee & g_change], 16 [g_coffee & g_change]	N/A	N/A	N/A
16	b_coffee:off, b_tea:off / s_coffee:on, s_tea:on / change:on	2% / 5, 16	3% / 5, 16	10% / 0, 16 [drop>0JPY]	N/A	N/A	25% / 10, 8 [drop=0JPY]	30% / 10, 14 [drop=0JPY]	30% / 10, 12 [drop=0JPY]

1 In the names of states, the label b_coffee and b_tea express the on/off of the lights that indicate whether the buttons to buy a coffee and a tea are available or not, respectively. Also, the label s_coffee and s_tea express the on/off of the lights that indicate whether coffees and teas are sold out or not, respectively. The label change expresses the on/off of the light that indicates whether there are insufficient 10 JPY coins for change or not.

2 In guards, the variable drop expresses the current total of dropped coins. Also, the variable 10JPYcoins expresses the current number of dropped 10 JPY coins. The variable g_coffee and g_tea indicate whether coffees and teas are sold out or not after the event occurrence, respectively. The variable g_change indicates whether there are insufficient 10 JPY coins for change or not after the event occurrence.

Fig. 3 OPFPI of software to control a vending machine

different runs yield similar values. Also, the average improvement is about 42 %, showing that the algorithm achieves significant improvements.

Depending on the selected parameters, evaluation value may converge in different fashions. Figure 4 shows that, in our case, it converges very fast. Therefore, there is a little change for $t > 10$ and no change for $t > 240$. This also means that sufficiently good test cases can be found without performing too many iterations of Step 2 (ACO).

The average time to complete the optimized test case generation is about 21 seconds on a laptop with 2.5 GHz CPU and 2 G Bytes RAM; it is acceptable for practical applications even if it needs to be executed many times by a trial-and-error

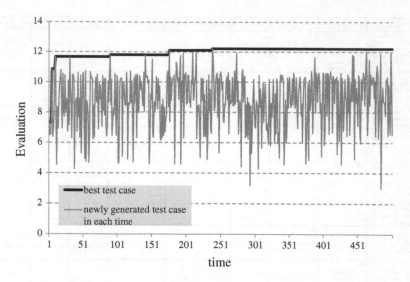

Fig. 4 Result of optimized test case generation

method. However, construction of an OPFPI model may be a relatively heavy task for test engineers. The quality of the OPFPI and the amount of time to construct it depend on test engineers' skill. When the SUT is developed by MDD (model-driven development) [6], the quality and time might improve due to the following reasons.

- Test engineers can get more familiar by gathering experiences while constructing models in requirements analysis and design processes. Also, they can use these models in order to efficiently construct a state machine for an OPFPI.
- Usage and fault distributions can be efficiently derived due to the traceability among models, source codes, and bug reports.

6 Conclusion

In this paper, a novel formal model called an OPFPI (operational profile with fault-proneness information) and a novel algorithm to generate optimized test cases from the OPFPI have been proposed in order to establish a testing technique that has a strength of both OPBT (operational profile-based testing) and FABT (fault analysis-based testing). The OPFPI has both usage distributions (i.e., information about how software will be used by users or other systems) and fault distributions (i.e., information about how software will include latent faults). It provides a good basis for generating test cases to effectively improve software reliability or gain a perspective of software reliability in a limited span of time for testing. The OPFPI includes the feasibility problem due to the use of guards (conditions to make specific transitions feasible); therefore, ACO (ant colony optimization) is employed in the algorithm to generate test cases that cover frequent and fault-prone transitions as comprehensively as possible. A test tool that implements the OPFPI model and executes the optimized test case generation has been developed, and the technique has been applied to a non-trivial system. The obtained results indicate that significant improvement of test cases can be achieved in a short time. However, constructing an OPFPI of good quality with a small amount of effort is a challenge, and MDD might be a key to tackle this problem.

In future study, we intend to consider a technique that enables inexperienced test engineers to effectively construct an OPFPI by using MDD tools, and we plan to extend the OPFPI and the test case generation algorithm in order to achieve other test tactics and perform further evaluations. Furthermore, the approach can also be extended to support traits such as hierarchy, modularity and concurrency in an OPFPI so that the efficiency can be increased while working with larger real-world software.

Acknowledgments This work was supported by JSPS KAKENHI Grant Number 23700038.

References

1. Beizer, B.: Software Testing Techniques, 2nd edn. Van Nostrand Reinhold, New York (1990)
2. Chruscielski, K., Tian, J.: An operational profile for the cartridge support software. In: Proceedings of 8th International Symposium on Software Reliability Engineering, pp. 203–212 (1997)
3. Cotroneo, D., Pietrantuono, R., Russo, S.: Combining operational and debug testing for improving reliability. IEEE Trans. Reliab. **62**(2), 408–423 (2013)
4. Doungsa-ard, C., Dahal, K., Hossain, A., Suwannasart, T.: Test data generation from UML state machine diagrams using GAs. In: Proceedings of International Conference on Software Engineering Advances, p. 47 (2007)
5. Eski, S., Buzluca, F.: An empirical study on object-oriented metrics and software evolution in order to reduce testing costs by predicting change-prone classes. In: Proceedings of International Conference on Software Testing, Verification and Validation Workshops, pp. 566–571 (2011)
6. Frankel, D.S.: Model Driven Architecture: Applying MDA to Enterprise Computing. Wiley, New York (2003)
7. Hartmann, H., Bokkerink, J., Ronteltap, V.: How to reduce your test process with 30 %—the application of operational profiles at Philips medical systems. In: Supplementary Proceedings of 17th International Symposium on Software Reliability Engineering. CD-ROM (2006)
8. Kalaji, A., Hierons, R.M., Swift, S.: Generating feasible transition paths for testing from an extended finite state machine (EFSM). In: Proceedings of International Conference on Software Testing Verification and Validation, pp. 230–239 (2009)
9. Musa, J.D.: The operational profile. Reliab. Maintenance Complex Syst. NATO ASI Ser. F Comput. Syst. Sci. **154**, 333–344 (1996)
10. Poore, J.H., Walton, G.H., Whittaker, J.A.: A constraint-based approach to the representation of software usage models. Inf. Softw. Technol. **42**(12), 825–833 (2000)
11. Rook, P.: Software Reliability Handbook. Elsevier Science, New York (1990)
12. Srivastava, P.R., Bahy, K.: Automated software testing using metahurestic technique based on an ant colony optimization. In: Proceedings of International Symposium on Electronic System Design, pp. 235–240 (2010)
13. Whittaker, J.A., Arbon, J., Carollo, J.: How Google Tests Software. Addison-Wesley Professional, Westford (2012)
14. Whittaker, J.A., Thomason, M.G.: A Markov chain model for statistical software testing. IEEE Trans. Softw. Eng. **20**(10), 812–824 (1994)

Design Pattern for Self-adaptive RTE Systems Monitoring

Mouna Ben Said, Yessine Hadj Kacem, Mickaël Kerboeuf, Nader Ben Amor and Mohamed Abid

Abstract Approaches for the development of self-adaptive real-time embedded (RTE) systems are numerous. However, there is still a lack of generic and reusable design which fits different systems and alleviate the designer task. Design patterns represent a promising solution to get fast and reusable design. Unfortunately, patterns dealing with self-adaptive RTE systems development are still not well tackled in the literature. The general structure of self-adaptive RTE systems is based on a MAPE loop which is composed of four basic adaptation processes: Monitor, Analyze, Plan, and Execute. In this paper, we define patterns for the monitoring and analyzing processes through the generalization of relevant existing adaptation approaches to improve their accessibility to new adaptive systems developers. To evaluate the work, the proposed patterns are applied to a relevant existing cross-layer adaptation framework.

Keywords Self-adaptive system · Real-time embedded system · Design patterns · Monitoring and analyzing

M. Ben Said (✉) · Y. Hadj Kacem · N. Ben Amor · M. Abid
ENIS, CES Laboratory, University of Sfax, Soukra Km 3,5 B.P.: 1173-3000,
Sfax, Tunisia
e-mail: mouna.ben-said@ccslab.org

Y. Hadj Kacem
e-mail: yessine.hadjkacem@ceslab.org

N. Ben Amor
e-mail: nader.ben-amor@ceslab.org

M. Abid
e-mail: mohamed.abid@ceslab.org

M. Kerboeuf
Lab-STICC, MOCS Team, University of Brest, Brest, France
e-mail: kerboeuf@univ-brest.fr

© Springer International Publishing Switzerland 2015
R. Lee (ed.), *Software Engineering Research, Management and Applications*,
Studies in Computational Intelligence 578, DOI 10.1007/978-3-319-11265-7_3

1 Introduction

The development of self-adaptive embedded systems is a costly and time-consuming process due to their complexity, continuously variable environment as well as the lack of reusable designs and development tools [1, 2]. Therefore, there is a high demand for alleviating the designer task and reducing cost and Time To Market by decreasing system complexity and automating the process management. At this moment, a question needs to be answered: How to exploit software engineering approaches to provide solutions that rapidly transform an ordinary embedded system into a self-adaptive one?

Many approaches for the development of self-adaptive RTE systems have been proposed in the literature [2, 3]. They are based on software engineering approaches, typically the Model Driven Engineering (MDE) which is well appropriate to embedded systems and permits to ease their design by avoiding dealing with technical details. However, there is still a lack of reusable design that is target independent and sufficiently generic to fit different systems and fasten the designer task. To address this problem, we propose to use design patterns which are effective for design reusability and quickness [4]. Patterns that are particularly intended for the embedded systems domain are still under-explored. Existing solutions are mostly dedicated to desktop and distributed systems and ignore important constraints of RTE systems, essentially the real-time constraint.

Our aim is to propose patterns for the development of self-adaptive RTE systems. These patterns provide a generic specification that guides experts of the domain to build their own adaptive systems giving them the freedom to insert additional information that have not been defined in the design. The general structure of self-adaptive RTE systems is based on a MAPE loop [5] which is composed of sensors, effectors and four basic adaptation processes which are Monitoring, Analyzing, Planning and Executing. Sensors are responsible for collecting data on the system context status. The monitoring process decides about relevant context changes based on the collected data and triggers change events. The analyzing process examines change events to detect if an adaptation is required. The planning process produces a decision specifying the system elements to change and how to change them in order to respect the system constraints. The executing module maps the decision actions to effectors which are related to the system adaptable elements and responsible for applying adaptation actions to them.

In this paper, we propose patterns for the monitoring and analyzing processes. We survey relevant research works related to adaptation in order to construct a unified terminology of adaptive systems that we use to develop our design patterns. These patterns are to be integrated in a model-based approach for automatic pattern-based re-factoring of the system design model to generate a dynamically adaptive system model. The patterns are described using Unified Modeling Language (UML) models that are annotated with the UML/MARTE (Modeling and Analysis of Real-Time and Embedded Systems) [6] profile stereotypes. The MARTE profile offers a rich terminology for the specification and analysis of RTE systems.

The present paper is organized as follows. In Sect. 2, we give the description of the proposed patterns. We illustrate in Sect. 3 the utilization and effectiveness of the patterns through a case study of dynamically adaptive RTE system running object tracking application. Section 4 provides a brief discussion of related work. To conclude, in Sect. 5, the suggested patterns are briefly outlined and future work is given.

2 Patterns Description

As we have already mentioned, in this paper we focus on the two first processes of the adaptation loop; monitoring and analyzing. The proposed patterns follow the [7] pattern template. In this paper, we give details of six fundamental fields which are the pattern name, problem, intent, context, motivation and solution. To describe our patterns, we use standard UML diagrams annotated with the UML/MARTE profile stereotypes. We explain for each pattern both structural and behavioral views using the class and sequence diagrams, respectively. These diagrams are the result of the abstractions and generalization of many relevant adaptation-related works [8–15].

2.1 RTE Monitor Pattern

Name: RTE Monitor
Problem:
The problem treated by this pattern is the detection of irregular status of an RTE system which results from relevant variation of a given internal or external context element.
Intent:
The *RTE Monitor* pattern permits the continuous control of the status of one RTE system property in order to detect relevant changes and trigger events. It takes into consideration the system stability issue by minimizing events trigger through the selection of only important context variations. It also handles concurrency and real-time features relative to the control operations.
Context:
This pattern is used in the first step of development of a self-adaptive RTE system. The designer has to define the system to adapt and his adaptation requirements by answering the «what to monitor?» question.
Motivation :
The starting point that triggers adaptation mechanism in an RTE system is the context variation detection. Therefore, to be self-adaptive, an RTE system first needs to integrate monitoring modules that permit to continuously control and update the status of its execution context. Additionally, the execution context of an RTE system is very fluctuant so that context variation detection risks to be very frequent. Therefore, in order to have a stable adaptive system with the minimum of reconfigurations, a monitoring step is required to restrict the number of treated changes by approving only relevant ones.

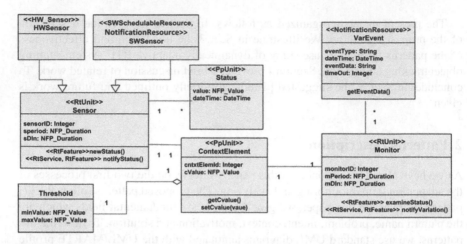

Fig. 1 Structural view of the RTE monitor pattern

Solution:

This pattern represents a monitoring process that permits the observation of the status of one RTE system context property.

Structural view:

Figure 1 shows the class diagram that explains the structural view of the pattern.

Participants:

- *ContextElement* represents an internal or external property of the system which is observed by the Monitor, such as CPU load, battery life and network bandwidth. It is a passive unit that carries information about the status of a system property and is concurrently accessed by the Sensor and Monitor. It is thus stereotyped «PpUnit». It specifies its concurrency policy through the *concPolicy* attribute (sequential, guarded or concurrent).

- *Sensors* are responsible for data collection about the status of a *ContextElement*. A *Sensor* is associated with each *ContextElement* and provides measures of its status. We classify sensors into two categories: a hardware sensor, stereotyped «HW_sensor», represents a hardware device providing a measure of a physical quantity and converting it into a signal. A software sensor is defined by a software task running concurrently on the system to measure a system property, such as CPU usage, and notify a *Monitor*. It is thus stereotyped «SwSchedulableResource» and «NotificationResource». A notification resource is a software synchronization resource used to notify events. To keep events history, the notified occurrences can be memorized in a buffer by setting the policy attribute to Memorized.

Since we are in the context of real-time and embedded systems, two basic issues have to be taken into consideration: the concurrency and the real-time features. In order to handle these issues, we annotate the active classes by the HLAM «RtUnit» stereotype [6] indicating that it is a real time unit. An RtUnit is an autonomous execution resource that may own one or more schedulable resources and one or several behaviors. It is also capable of handling different incoming messages at the same time without worrying about concurrency issues thanks to its own concurrency and behavior controller. It owns a message queue permitting to save messages it receives. Messages can represent operation calls, signal occurrences or data receptions. A message can be used to trigger the execution of a behavior owned by the real-time unit. A sensor is an active class that we annotate «RtUnit».

- *Status* stores the measures realized by the sensor in order to keep track of context information history which is important to determine the trends of the *ContextElement* variations [13] and consequently to improve predictions. Status indicates for each measure the date and time and the value. This latter is typed *NFPValue* which has different attributes that permit to precisely specify NFP values, such as statistical qualifier, precision and source.

- The *Monitor* is an «RtUnit» that is associated to a *ContextElement*. It examines sensing data using minimum and maximum values stored in a *Threshold* that is relative to the considered *ContextElement* to decide if a significant variation has occurred or a certain threshold has been exceeded. Threshold may represent either interval limits indicating a regular status or allowed variation margins to be used to decide about the variation relevance. If a variation is relevant, the *Monitor* generates a variation event, stereotyped «NotificationResource».

For the sake of system stability when self-adapting, it is recommended to define an adaptation period in order to manage the adaptation mechanism occurrence. Commonly, this period is equal to a defined number *Ne* of application iterations [11]. Every period, the monitoring process starts a control session and then the adaptation cycle is executed. Therefore, the operations executed by the sensing and monitoring processes need to specify their occurrence kind (such as periodic, aperiodic and sporadic). Moreover, in order to respect the real-time constraint, these operations have deadlines that they are asked to meet. In order to model these real-time properties, we annotate the sensing and monitoring methods with the «RtFeature» stereotype which has the occKind, relDl and absDl (for occurrence kind, relative deadline and absolute deadline, respectively) attributes. In order to specify additional attributes for real-time constraints of these operations, we use the «RtService» stereotype. It permits to manage the execution priority of a real-time service by the specification of the execution kind (exeKind attribute) which can be either deferred, remoteImmediate or localImmediate.

Behavioral view:
Figure 2 shows the UML sequence diagram presenting the execution scenario of the *RTE Monitor* pattern by showing the communication between the different objects forming it. The monitoring process starts by *Sensor* which periodically delivers a new measure of the status of the supervised *ContextElement* then it notifies the *Monitor*. The *NotifyStatus()* method execution kind is *localImmediate* in order to be immediately executed by the *Sensor*. The *Monitor* receives the new measure and updates the current status value of the *ContextElement*. Then it examines the

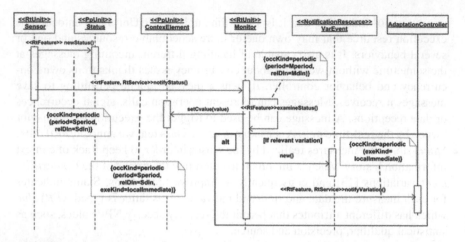

Fig. 2 Behavioral view of the RTE monitor pattern

new status to decide about the relevance of the change. It can use thresholds to verify whether the measure is in the interval delimited by min and max values. The negative case indicates an irregular state causing the *Monitor* to generate a variation event and send it to the *Adaptation Controller* of the system through its *notifyVariation()* method in order to be processed and decided upon. This method occurrence is aperiodic since its execution depends on the verification result of the *examineStatus()* method. However, when executed, it has the highest priority, thus having its execution kind set to *localImmediate*.

2.2 RTE Analyzer Pattern

Name: RTE Analyzer
Problem: having the status of an RTE system context, the *RTE Analyzer* pattern responds to the question «Does an adaptation need to be applied?»
Intent:
This pattern permits the verification of constraints meeting of an RTE system and then asks for adaptation if needed. It contributes to providing a stable adaptive system by minimizing adaptation requests. It handles concurrency and real-time features relative to the control operations.
Context:
This pattern is used when designing a self-adaptive RTE system, specifically when information about changes in the system context is available and system constraints are defined.
Motivation:
A change in the execution context does not necessarily affect the functioning of the system, i.e. violate system constraints, thus not requiring an adaptation. Therefore,

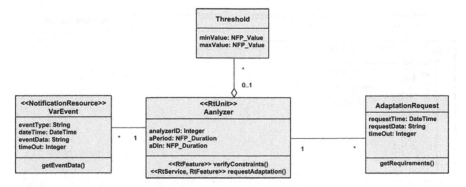

Fig. 3 Structural view of the RTE analyzer pattern

a verification step is needed in order to avoid useless adaptations.

Solution:

Structural view :

Figure 3 shows the class diagram relative to the structural view of the *RTE Analyzer* pattern.

Participants:

- The *Analyzer* is responsible for the verification of the system constraints meeting. It processes a *Variation Event* that occurs to the system to decide whether an adaptation action is required or not. It is thus an active class stereotyped «RtUnit». It has an analysis method, *verifyConstraints()*, which generally executes a constraint miss test. The miss-test may require *thresholds*. The *requireAdaptation()* method generates an *Adaptation Request*.
- An *AdaptationRequest* carries request data indicating the analysis results such as the source of constraint violation. It has a timeout to be considered when treated.

Behavioral view:

The behavior of the Analyzer pattern is depicted by the UML sequence diagram in Fig. 4. Having variation events received in its message queue, the Analyzer treats them in a loop. It asks for event data if the event is still valid, i.e. its timeout is not achieved. Then it uses the collected data to verify the system constraints meeting through the *verifyConstraints()* method. Since events occurrence is aperiodic, this method's occurrence kind is aperiodic too. If constraints are not met, the analyzer asks for adaptation by sending an *AdaptationRequest* to the *Adatation Controller* of the system. For more clarity, we can cite an example of real scenario: when a task entry event occurs in the system, the Analyzer performs a schedulability test to verify the real-time constraint meeting. If tasks' deadlines are not met, it asks for adaptation by generating an adaptation request carrying new context data.

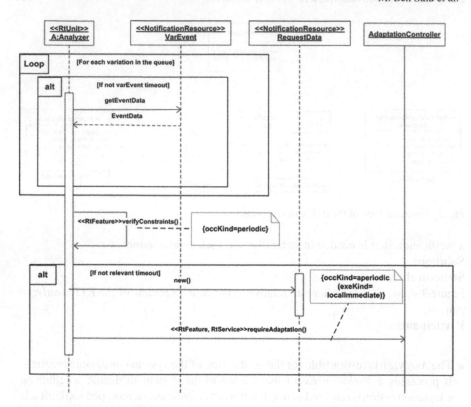

Fig. 4 Behavioral view of the RTE analyzer pattern

3 Example of Pattern Application

This section permits to illustrate the utilization of the proposed patterns through a case study of a dynamically adaptive object tracking application proposed in [11]. This work has been developed at a low level, thus lacking a high abstraction level modeling step, which is the case of most state of the art works in the field. Consequently, to apply our patterns, we start by providing a table of correspondence between adaptation concepts offered by the patterns and those considered in the application example in order to evaluate the generic aspect of the solutions. Then, we present a pattern-based design to show how to append the patterns to the self-adaptive system structure. Since this paper focuses only on the monitoring and analyzing processes, this section is thus limited to the adaptation features related to these two modules.

3.1 Application Description

The application presented in [11] consists in a self-adaptive object tracking application implemented on an FPGA-based smart-camera. The application is composed of

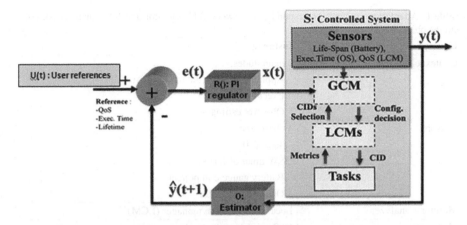

Fig. 5 Global structure of the closed-loop self-adaptive system proposed in [11]

10 tasks which can be implemented in HW or in SW. An electric toy train tracking scenario is proposed to illustrate the system self-adaptivity. The scenario contains various events provoking configuration decisions. The goal is to design an embedded system able to respect a constraint while optimizing secondary magnitudes. In this case study, the regulated magnitude is the QoS indicating the tracking accuracy and the optimized ones are power and execution time.

This self-adaptive system is developed based on a cross-layer adaptation approach for self-adaptive RTE systems development. Authors propose a hierarchy of local and global configuration managers (LCM/GCM), as depicted in Fig. 5, to separately deal with application-specific and application-independent reconfigurations, respectively. The LCM is responsible for local application algorithmic reconfigurations. An LCM is defined for each application. The GCM is responsible for architectural reconfigurations which consist in tasks migration from software to hardware on a multiprocessor heterogeneous architecture. Only one GCM is defined for the whole system.

3.2 Adaptation Concepts Correspondence

The adaptation features correspondence is summarized in Table 1.

The train tracking application considers three *ContextElements*, the QoS metric indicating the tracking error, the execution time and the battery life. It uses three corresponding sensors and a fourth one providing estimations of the context behavior. A LCM compares user-defined references with observed magnitudes values to detect irregular status. It then asks for correction by sending a preliminary decision of application's algorithmic configuration to the GCM. This latter is the responsible for the global reconfiguration decision, taking into account the received LCM local decision. A configuration period Ne is defined to control the periodic execution

Table 1 Adaptation concepts correspondence between RTE monitor and RTE analyzer patterns and the object tracking application

Concept	Instance
Context elements	3 magnitudes: Application QoS (the tracking error) Execution time Observer estimator
Sensors	4 observers: Task T10 SW timer of RTOS Battery gauge component Observer estimator
Monitor + analyzer	Local configuration manager (LCM)
Monitoring thresholds—irregular status	QoS reference (the tracking maximum error) is set to 10 % and reduced to 2 % within the critical area to guaranty good reactivity Task T10 provides the LCM with the application QoS metric (error between prediction and object position A value close to 0 but lower than the reference (10 %) means a very high tracking quality –> can be relaxed by reducing the application speed A value higher than the reference –> the application rate must be increased with a faster configuration
Decision maker	Global configuration manager (GCM)
Adaptation request (request data)	An Algorithmic configuration as a first local decision sent by the LCM to the GCM mailbox
System stability and avoidance of reconfiguration	Proportional Integrator (PI) regulator (coefficients: $kp = 0, 25; ki = 0, 25$), and least mean square (LMS) observer (coefficient $kL = 2^{21}$)
Adaptation period	Configuration period Ne is set to 1, which means (that a new configuration is evaluated after each application iteration)

of the adaptation mechanism. This scenario is very similar, in both structure and behavior, to the *RTE Monitor* pattern in conjunction with the *RTE Analyzer*. The LCM encompasses both monitoring and analyzing modules. The three magnitudes are the observed *ContextElements* whose measured values, with a set of associated thresholds, form the input of the LCM. Its output is an adaptation request sent to the Decision Maker (the GCM) where the request data is an Algorithmic configuration.

This equivalence study permits to show the important degree of similarity between the proposed patterns and the application case study. That proves the efficiency of the solution in providing a generalized and high abstracted design permitting to cover most RTE adaptation features design.

3.3 Patterns Application

The general structure of the closed-loop self-adaptive system is illustrated in Fig. 5. It is composed of 4 basic elements:

- The controlled system S is composed of configuration managers, a LCM for the tracking application and a GCM, a set of tasks and sensors observing the controlled magnitude y(t).
- The control function R is a proportional and integrator regulator (PI) permitting to handle system stability and avoidance of reconfiguration.
- The system observer O calculates estimates of controlled magnitude for the next time slot in order to predict its evolution to anticipate the right reconfiguration decision. This model-based estimator permits rapid, accurate and costless estimation of system behavior that replaces new sensors' measures when they are delayed or even not available.
- The user references u(t) represent values or thresholds relative to considered magnitudes, that are defined by the designer to be used by the control mechanism. For instance, a reference can be a lifetime threshold that will be compared to a value provided by the battery gauge component, or an application QoS constraint that will be compared to the QoS value provided by the LCM of the application.

The corresponding high-level pattern-based design is depicted in Fig. 6. We define three instances of the *RTE Monitor* pattern, one for each controlled magnitude. Then, one instance of the *RTE Analyzer* is defined having as input the *VariationEvents* generated by the monitors. The analyzer is then linked to the rest of the system components which are responsible for the remaining adaptation processes.

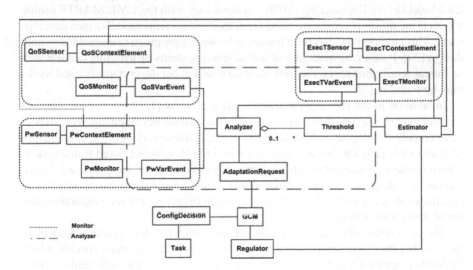

Fig. 6 Patterns application

3.4 Discussion

Using the high abstraction level pattern-based modeling for the development of a RTE system presents advantages upon the low-level development approach. Instantiating the patterns for the target system has simplified the design by hiding internal functional details of system elements and lowering the system model size. This follows from the fact that the patterns operate at a high abstraction level to cover adaptation features considered by the low-level system design. In addition, the pattern-based design permitted us to promote modularity and thus flexibility of the design by providing a modular structure for the LCM component.

4 Related Work

The development of self-adaptive systems has begun several years ago and proposed approaches and techniques have evolved with the evolution of technology, system architectures, applications, user requirements and environment constraints. In this section, we limit our review to relevant research works on adaptation of embedded systems, which are the most useful for the definition of our adaptation patterns.

Self-adaptation in embedded systems has been well tackled in the literature. Several approaches have been based on low-level development process integrating adaptation techniques in the classic System on Chip co-design flow [8–11]. Later, the development of adaptive systems at a low-level has become a tedious task due to the growing complexity of modern systems and dedicated applications. Designers have then resorted to high abstraction level approaches [16–19], typically based on the Model Driven Engineering (MDE) methodology with the UML/MARTE profile which is the most upcoming standard for embedded systems model-driven development. This methodology has been proven to be well appropriate to embedded system design [20]. It eases the modeling of self-adaptive systems by avoiding dealing with technical details thus promoting reusability. Details about the previously cited works may be found in [3].

In addition to the above approaches, several projects [12] have been realized in the literature to help guide and ease self-adaptive systems development. Researchers have equally developed middleware [13, 21, 22], frameworks [14, 23], languages [24] and tools [25] for this aim. For example, in [14], a model based framework to automate the development of self-adaptive embedded systems was proposed. It consists in extending an original system model to a self-adaptive one based on a formal specification of the embedded system, the events triggering the reconfiguration and the reconfiguration requirements.

The cited approaches permit to facilitate and fasten the development of adaptive systems but they have some limitations. They are generally domain specific which limits their applicability for diverse systems. They are also not sufficiently generic since they tackle specific adaptation problems, which consequently compromises their reusability as well as their ability to adapt to new system requirements and constraints. Additionally, most of them focus on the software side adaptation while ignoring the hardware side which is essential in the embedded systems design.

The use of design patterns seems to be a promising solution for the above problems. A design pattern gives a higher abstraction view of a commonly recurring problem and promotes the reusability and extensibility of the design. Research works dealing with pattern-based adaptation are limited. We distinguish two groups of pattern-related works. The first group was dedicated to the organization of adaptation functions. In [26], Puviani et al. proposed a taxonomy for self-adaptation patterns at both component and ensemble levels. The basic components that may compose self-adaptation patterns represent the component level and the mechanisms by which components can be composed into ensembles represent the ensemble level. Authors of [27] proposed patterns to decentralize multiple adaptation loops in large and complex self-adaptive systems.

The second group of works rather dealt with the internals of adaptation modules [4, 28, 29]. Gamaa and colleagues [4] proposed design patterns to specify the behavior to dynamically reconfiguring software architectures. Four types of architectures were considered: master/slave, centralized, server/client, and decentralized. In [28], a set of patterns for the development of adaptive middleware has been proposed. In [29] authors proposed a set of patterns permitting to adapt distributed networked systems to satisfy the requirements and constraints which arise at runtime. Three categories of patterns were presented: monitoring, decision-making and reconfiguration activities. These patterns are beneficial for the development of adaptive systems in different domains. However, they do not deal with RTE systems constraints which represent a key issue for the real-time and embedded systems domain. Also, they only address the software part of the system and are most appropriate for distributed systems.

5 Conclusion and Future Work

In this paper, we proposed a generic design of self-adaptive RTE systems based on an adaptation loop composed of four adaptation processes, monitoring, analyzing, deciding and acting, accompanied with sensors and effectors. We detailed the internals of the two first adaptation processes. We proposed design patterns giving generic solutions for the monitoring of an RTE system context and the verification of its constraints meeting to decide about the need of an adaptation action. Both patterns deal with the system stability issue by minimizing events trigger through the selection of only important context variations. In addition, they handle concurrency and real time features of adaptation operations. The proposed solution permits to guide adaptive systems designers decrease the complexity of their heavy job and fasten the development of such systems by promoting reusability of the design.

The proposed patterns were applied to a RTE system running a self-adaptive object tracking application implemented on an FPGA-based smart-camera. The case study shows the effectiveness of using high abstraction level pattern-based modeling for the development of a RTE system. It promotes modularity and flexibility of the design by providing a modular structure.

We plan in future work to complete the modules of the adaptation loop. Indeed, our goal is to integrate the proposed patterns in an MDE-based approach for the automatic generation of self-adaptive RTE systems.

References

1. Kofod-Petersen, A., Mikalsen, M.: Context: representation and reasoning. Representing and reasoning about context in a mobile environment. Revue d'Intell. Artif. **19**(3), 479–498 (2005)
2. Salehie, M., Tahvildari, L.: Self-adaptive software: landscape and research challenges. ACM Trans. Auton. Adapt. Syst. **4**(2), 14:1–14:42 (2009)
3. Said, M.B., Kacem, Y.H., Amor, N.B., Abid, M.: High level design of adaptive real-time embedded systems: a survey. In: International Conference on Model-Driven Engineering and Software Development—MODELSWARD 2013, pp. 341–350. 19–21 Feb 2013
4. Gamma, E., Helm, R., Johnson, R., Vlissides, J.: Design Patterns: Elements of Reusable Object-Oriented Software. Addison-Wesley Longman Publishing Co., Inc., Boston (1995)
5. Kephart, J.O., Chess, D.M.: The vision of autonomic computing. Computer **36**(1), 41–50 (2003)
6. A UML Profile for MARTE: Modeling and Analysis of Real-Time Embedded Systems, ptc/2011-06-02. Object Management Group, Needham (2011)
7. Buschmann, F., Meunier, R., Rohnert, H., Sommerlad, P., Stal, M.: Pattern-Oriented Software Architecture: A System of Patterns. John Wiley & Sons Inc, New York (1996)
8. Yuan, W., Nahrstedt, K.: Energy-efficient cpu scheduling for multimedia applications. ACM Trans. Comput. Syst. **24**(3), 292–331 (2006)
9. Vardhan, V., Yuan, W., Harris III, A.F., Adve, S.V., Kravets, R., Nahrstedt, K., Sachs, D.G., Jones, D.L.: Grace-2: integrating fine-grained application adaptation with global adaptation for saving energy. IJES, **4**, 152–169 (2009)
10. Ye, L., Diguet, J.-P., Gogniat, G.: Rapid application development on multi-processor reconfigurable systems. In: FPL, pp. 285–290, 2010
11. Diguet, J.P., Eustache, Y., Gogniat, G.: Closed-loop-based self-adaptive hardware/software-embedded systems: design methodology and smart cam case study. ACM Trans. Embed. Comput. Syst. **10**(3), 38:1–38:28 (2011)
12. Nishanth, S., Kinnebrew, J.S., Koutsoukas, X.D., Chenyang, L., Schmidt, D.C., Biswas, G.: An integrated planning and adaptive resource management architecture for distributed real-time embedded systems. IEEE Trans. Comput. **58**(11), 1485–1499 (2009)
13. Mikalsen, M., Paspallis, N., Floch, J., Stav, E., Papadopoulos, G.A., Chimaris, A.: Distributed context management in a mobility and adaptation enabling middleware (madam). In: SAC, pp. 733–734, 2006
14. Li, T.: Model-based self-adaptive embedded programs with temporal logic specifications. In: Proceedings of the Sixth International Conference on Quality Software, QSIC06, pp. 151–158. IEEE Computer Society, Washington, DC, USA, 2006
15. Andersson, J., Lemos, R., Malek, S., Weyns, D.: Software Engineering for Self-Adaptive Systems. Chapter Modeling Dimensions of Self-Adaptive Software Systems, pp. 27–47. Springer, Berlin (2009)
16. Quadri, I.R., Yu, H., Gamatié, A., Rutten, E., Meftali, S., Dekeyser, J.-L.: Targeting reconfigurable FPGA based SoCs using the MARTE UML profile: from high abstraction levels to code generation. Special Issue on Reconfigurable and Multicore Embedded Systems. Int. J. Embed. Syst. **4**(3/4), 204–224 (2010)
17. Vidal, J., de Lamotte, F., Gogniat, G., Diguet, J.-P., Soulard, P.: Uml design for dynamically reconfigurable multiprocessor embedded systems. In: Proceedings of the Conference on

Design, Automation and Test in Europe, DATE 10, pp. 1195–1200. European Design and Automation Association, 3001 Leuven, Belgium, Belgium, 2010
18. Krichen, F., Hamid, B., Zalila, B., Jmaiel, M.: Towards a model-based approach for reconfigurable dre systems. In: ECSA, pp. 295–302, 2011
19. Zhang, J., Cheng, B.H.C.: Model-based development of dynamically adaptive software. In: Proceedings of the 28th International Conference on Software Engineering, ICSE '06, pp. 371–380. ACM, New York, NY, USA, 2006
20. Gogniat, G., Vidal, J., Ye, L., Crenne, J., Guillet, S., de Lamotte, F., Diguet, J.-P., Bomel, P.: Self-reconfigurable embedded systems: from modeling to implementation. In: ERSA, pp. 84–96, 2010
21. Capra, L., Emmerich, W., Mascolo, C.: Carisma: context-aware reflective middleware system for mobile applications. IEEE Trans. Softw. Eng. **29**(10), 929–945 (2003)
22. Sadjadi, S.M., McKinley, P.K.: A survey of adaptive middleware. Technical Report MSU-CSE-03-35, Department of Computer Science, Michigan State University, East Lansing, Michigan, Dec 2003
23. Yuana, W., Nahrstedta, K., Advea, S.V., Jonesb, D.L.: Design and evaluation of a cross-layer adaptation framework for mobile multimedia systems. In: Proceedings of SPIE 5019, Multimedia Computing and Networking, 2003
24. Vogel, T., Giese, H.: Model-driven engineering of adaptation engines for self-adaptive software: executable runtime megamodels. Technical Report 66, Hasso Plattner Institute for Software Systems Engineering, University of Potsdam, Germany, 4, 2013
25. Bencomo, N., Grace, P., Flores, C., Hughes, D., Blair, G.: Genie: supporting the model driven development of reflective, component-based adaptive systems. In: Proceedings of the 30th international conference on Software engineering, ICSE '08, pp. 811–814. ACM, New York, NY, USA, 2008
26. Puviani, M., Cabri, G., Zambonelli, F.: A taxonomy of architectural patterns for self-adaptive systems. In: Proceedings of the International C* Conference on Computer Science and Software Engineering, C3S2E '13, pp. 77–85. ACM, New York, NY, USA, 2013
27. Weyns, D., Schmerl, B., Grassi, V., Malek, S., Mirandola, R., Prehofer, C., Wuttke, J., Andersson, J., Giese, H., Göschka, K.: On patterns for decentralized control in self-adaptive systems. In: Lemos, R., Giese, H., Müller, H.A., Shaw, M. (eds.) Software Engineering for Self-Adaptive Systems II. Lecture Notes in Computer Scienedsce, vol. 7475, pp. 76–107. Springer, Berlin (2013)
28. Schmidt, D., Stal, M., Rohnert, H., Buschmann, F.: Pattern-oriented software architecture. In: Volume 2: Patterns for Concurrent and Networked Objects. Wiley, 2000
29. Ramirez, A.J., Cheng, B.H.C.: Design patterns for developing dynamically adaptive systems. In: Proceedings of the 2010 ICSE Workshop on Software Engineering for Adaptive and Self-Managing Systems, SEAMS '10, pp. 49–58. ACM., New York, NY, USA, 2010

Changing the Software Engineering Education: A Report from Current Situation in Mexico

I. García, C. Pacheco and J. Calvo-Manzano

Abstract Nowadays, in Mexico software engineering education has two problems when satisfying software industry necessities: the quantity of young and skilled students and the quality of their formation. In this sense, it is necessary to improve education at the undergraduate level. We have identified five malfunctions in the current situation of software engineering education. The correction of these problems requires the design of a module-oriented curriculum, creation of networks with local industry, federal and state funding, and alternative educational paradigms for software engineering. This paper shows the strategy and federal programs established to modernize the software engineering curriculum.

Keywords Software engineering education · Educational paradigms in software engineering · Federal programs · Funding opportunities

1 Introduction

In Mexico, the software industry is still immersed in a process of development and growth. Mexico has an Information Technology (IT) expenditure level of 1.4 % in relation to the Gross Domestic Product (GDP) and is located in 19th place at worldwide level. This is a poor expenditure in comparison to a world average of 4.3 % for countries in the Organisation for Economic Cooperation and Development (OECD) and to 5.5 % in the USA [25]. The participation of software industry in IT is low; however, it is important to take into account two issues: firstly, the software industry

I. García (✉) · C. Pacheco
Postgraduate Division, Universidad Tecnologica de la Mixteca,
68020 Huajuapan de León, Oaxaca, Mexico
e-mail: ivan@mixteco.utm.mx

C. Pacheco
e-mail: leninca@mixteco.utm.mx

Calvo-Manzano, J.
Universidad Politecnica de Madrid, Computer Science School, 28060 Madrid, Spain
e-mail: jacalvo@fi.upm.es

© Springer International Publishing Switzerland 2015   43
R. Lee (ed.), *Software Engineering Research, Management and Applications*,
Studies in Computational Intelligence 578, DOI 10.1007/978-3-319-11265-7_4

only considers packaged software and not the customized one, this is registered in the IT services segment. The main reason for this is, in part, because it is considered an activity specific to user necessities and it is not produced in large quantities as is packaged software [23]. Secondly, the values of IT services segments are bigger than the software ones because they register, besides the customized software, other services such as consultancy and training. Following this reasoning, the total of software production in Mexico is represented by 29.4 % of the total for packaged software, 8.0 % of the total for customized software, and 62.6 % for production and personal consumption. The interesting point is the fact that an increasing proportion of worldwide software production is performed in "developing countries". India and Ireland represent two successful cases of creation and growth of national industries based on exportation. It is obvious that the Mexican software industry is very young to be compared with both countries; however, the government has paid attention to promote initiatives related to improve it. Thus, the Secretariat of Economy and enterprise organisms designed the Software Industry Development Program (PROSOFT) [26] to promote the software industry and extend the IT market in Mexico. As a consequence, a new model was proposed for Mexican industry, MoProSoft [16], which is developed taking into account the better practices of models such as CMMI-SW [5], ISO 9000:2000 [22], PMBoK [20], among others. However, besides this effort, technology and education should be aligned to walk the same road.

Consequently, Sect. 2 examines the current situation of software engineering education through a survey of software engineering-related courses offered in the higher education universities in Mexico. Section 3 presents the remedy proposed to solve the malfunctions identified in the survey. Section 4 summarizes results obtained from the improvement efforts, and finally Sect. 5 summarizes our conclusions and future actions.

2 Malfunctions of Software Engineering in Mexico

We performed an analysis of software engineering courses offered in the academic year 2004 (see Table 1). Data were collected from the web site of courses offering listing and triptychs from most of the Mexican universities incorporated into the Public Education Secretariat of Mexico. We selected this year because of two crucial issues in Mexican software engineering education: firstly, an important restructuring of software engineering curricula was performed in 2004; and secondly, the appearance of the Mexican model MoProSoft in 2005 relied on a restructured curricula for software engineering at 2004.

A sample of 210 universities was included in the survey. 182 out of 210 universities offered software engineering-related courses, while 2 courses were offered as module programs by eight universities, and a sum of 292 software engineering-related courses were offered. Two major problems with the software engineering education in Mexico can be derived from the data shown in Table 1:

Table 1 Software engineering-related courses offered in 2004

Course title	No. of universities	No. of courses
Software engineering	74	82
Software architecture	6	6
Software project management	87	87
Software testing	1	1
Software integration	0	0
Component-oriented software development	0	0
Object-oriented software development	0	0
Patterns-based software design	0	0
Software modeling and analysis	67	67
Software enterprises management	0	0
Software metrics	1	1
Software projects development	67	67
Secure software development	0	0
Software verification and validation	0	0
Software engineering advanced topics	0	0
Business processes modeling	0	0
Software reference models	0	0
Software agile development	0	0
Software projects evaluation	10	10
Software quality	36	36
Software process assessment	0	0
Software product workshop	0	0
Software industry in the national context	0	0
Software marketing	0	0
Professional labor of software engineering	0	0
Software normativity	0	0
Formal specification for software design	1	1
Continuous improvement	0	0
Software requirements development	1	1
PSP	0	0
TSP	0	0
Six Sigma	0	0
Risk management for software projects	0	0
Software factories	0	0
MoProSoft	0	0
CMMI	0	0
Teach a single course	**60**	
Module program education	**8**	

- Assuming an average of 20 enrollments in each software engineering-related course, only some 5,840 students (292 courses multiplied by 20 enrollments per course) took these courses in the 2004 academic year. However, according to the Mexican Secretariat of Labor, only 25 % of these students had the skills that the software industry required in 2005 when MoProSoft was introduced in small-sized software enterprises. Thus, a real value of 1,460 students represented the shortage of manpower of software engineers.
- In relation to the skills, a more important malfunction is that most of the software engineering-related courses, 82 of 292, were "Software Engineering" which is more oriented to introductory topics, 87 courses on "Software Project Management" (70 courses were focused on general projects and only 17 were focused on software projects), 67 courses were "Software Modeling and Analysis" as a basis for software development, and a small number-36 courses-on "Software Quality" only addressed very limited knowledge.

As can be seen, in 2004 there were problems related to both quantity and quality of students. However, it is interesting to note that as late as 2008, an urgent call for software engineering education action was made by Watts Humphrey in an International Conference on Software Engineering held in Mexico. In this sense, in [11] Humphrey affirmed that the Mexican software industry faces a big challenge for growth; basically Mexico must integrate government-industry-academia support and should establish modern curricula to create a body of highly qualified engineers, and it must increase the courses related to knowledge and skill-based PSP [9] formation and performance-based TSP training [10]. It is true that the Mexican government attempts to support the software industry through the PROSOFT initiative, but without the correct economic resources destined to education (just 0.9 % of the GDP goes to graduate education), universities are limited in providing modern education. Furthermore, a reference model, like CMMI family models and MoProSoft, assumes by its nature that software engineers know how to estimate cost and schedule, conduct project planning, elicitate requirements, or undertake configuration management, for example, but this knowledge is not offered nor observed in Table 1.

In an effort to solve the detected problems, we identified five malfunctions in the current situation of Mexican software engineering education: inattention to software process, lack of awareness of software process quality, little concern to plan and control the software projects, a chasm between courses and software industry, and lack of specialization in specific skills. These malfunctions represented the basis for research which topics and courses should be developed to further strengthen the software engineering knowledge, and thereby increase the rate of enrollment in software engineering-related courses.

This research has been performed since 2004 when problems, analysis and derived malfunctions were obtained; and it has continued with careful analysis over 7 years with collaboration from industry and government departments. Further efforts performed to change software engineering education in Mexico are shown in Sect. 4 summarizing experiences working on software projects, serving on several software-related reviewing committees, and reviewing numerous Mexican software engineering-related curricula.

2.1 Inattention to Software Process

Normally a Mexican software engineering related-course focused on traditional (universal) process models (i.e. waterfall model and spiral model); with this established process definition students understood what they should do, what they could expect from their co-workers, and what they were expected to provide in return. This allowed them to focus on doing their job. However, students experienced difficulty in learning how to correctly use/select a process model within a complex and large software project. A "Software Reference Models" course would explain the importance of a complex software process where a process model (such as the spiral model) is a minimum part of a whole process. Under these 2004 circumstances, undergraduate students learned to work independently with little coordination and obtained a mistaken conception of software process because it was often viewed as overhead, that is, paper work that brings little benefit to justify the cost. However as software products become more complex, the software process becomes too complex to be tackled by individuals in a time and cost-efficient manner. While this malfunction can have many causes, as far as software engineering education is concerned, universities had not been paying enough attention to courses oriented to software process. To help turn things around, "Software Reference Models", "MoProSoft" and "CMMI" courses should receive the right amount of attention.

2.2 Lack of Awareness to Software Process Quality

We were able to observe a general lack of concern in software engineering-related courses about quality in software products and in software development process in overall. In a programming course (prior to any software engineering course), students learn to confuse testing with debugging. Regularly, they perform a lot of debugging to corroborate that their codes are sufficient to be delivered, but programs are not tested. Most of the programs that the students wrote during this analysis were only debugged but not tested, and rarely were they run for more than a few times. According to 2004 curricula, there is not too much time for testing.

Another problem observed is that an important number of Mexican universities include usability courses to support teaching of software quality topics. We determined an exaggerated focus on usability tests as a false measurement of software quality. It is common to hear that some students believe that "usability incorporates quality into the software process". This is a big problem, because this mentality carries over when graduates enter the job market. In this sense, a good change would be that courses such as "Software Process Improvement", "Verification and Validation", "Software Testing" and "Software Quality" receive greater attention in the curricula.

2.3 Little Concern to Plan and Control the Software Projects

To learn about the current scene, as an additional task, in 2011 we designed a modeling-based approach to support software process assessment in Mexican small software enterprises [7]. At the end of 2011 we used this assessment mechanism in 96 software enterprises which use MoProSoft as reference model. For this purpose, we carefully selected small enterprises which had recently hired at least five graduates in the last year. Our objective was to obtain a snapshot of graduates' education once that they were inserted in the industrial context. In reviewing the obtained results, it was common to find that 92 % of hired students were unable to produce a description of a common process for planning and control of software projects. But it was more worrisome to observe that 85 % of the enterprises' young engineers did not know how to estimate costs, generate schedules and budgets, establish measurement techniques, or measures for project monitoring and control. The latter makes clear that a "Software Project Management" course in 2004 was more focused on general projects (80.45 % of courses) and universities were not concerned about teaching how to plan and control software projects specifically. We detected that Mexican education in software engineering in 2004 was more centered on programming topics than on software project managerial skills.

2.4 Chasm Between Courses and Software Industry

By reviewing the courses offered in 2004 as shown in Table 1, we found that none of the courses emphasized connecting the knowledge gap to better prepare our students for the software enterprise context they may encounter in their roles as software engineers. In this sense, one of the most crucial problems in Mexican software industry is that software enterprises normally have to invest huge economic resources in the training of new workers due their knowledge deficiencies. In a real industrial environment, software engineers must effectively communicate with users, customers and domain experts in order to capture the correct requirements; estimate in an efficient manner the project costs and schedule; formulate appropriate plans; monitor and control the established plans; and take corrective actions when necessary. An important problem in software engineering education in Mexico is the gap between the courses offered in universities and the real necessities of the software industry. In this sense, curricula have to be constantly updated according to software demands in the national industry; but it is more important to relate courses with local software industry in order to prepare undergraduate students in a real working environment.

2.5 Lack of Specialization in Specific Skills

In this sense, through our study we were able to determine that most of the software engineering-related courses were based on expository classes which used the support of practical extracurricular exercises to complement the provided theoretical material. This "conventional" way of teaching promotes individualism in the classroom, and segregates skilled students from the rest. Thus, the observed problems with the 2004 curricula, shown in Table 1, were the following: (1) the offered courses were focused on general issues and software engineering basic skills, (2) the majority of the software engineering-related courses followed the expository classes teaching technique, promoting individualism among students, (3) no specific skills were included as a course or a topic within courses (specialized skills as PSP, TSP, Six Sigma, MoProSoft, or CMMI, for example); and more importantly, the problem that topics such as cost estimation, software metrics, or requirements development process were not included in the students' education.

3 A Suggested Rectification

After 7 years of changes, the Mexican software industry has received funding at the national level, through the Secretariat of Economy, to promote its internationalization. Government programs have been created to support this task with the collaboration of academics and IT specialists. Moreover, universities have updated their curricula for computer science and IT-education (where software engineering-related courses are included) and some of them have included, in 2006–2007, the "Software Engineering" program for undergraduate education. These efforts have been done to solve problems of shortage and low quality of software engineers by correcting the five identified malfunctions. This effort relies on the establishment and promotion of a modern educational plan to organize people and activities, a modernization of adaptable curricula according to industry necessities, and a feasible strategy for allocating resources.

3.1 Feasible Strategy for Allocating Resources

As a primordial strategy to promote the strengthening of IT and software industries, the Mexican government established in its National Plan for Development 2007–2012 [24] the objective of promoting productivity and competitiveness of the Mexican economy to achieve sustainable economic growth and to accelerate the creation of jobs to improve the quality of life of Mexicans. To accomplish the objective of increasing the country's competitiveness, the Subsecretariat of Industry and Commerce of the Secretariat of Economy launched in February 2008, Ten Guidelines to Increase Competitiveness, 2008–2012. The eighth guideline of this programme proposes to transform Mexico into the axis for distributing IT services and logistics, taking advantage of the geographical advantages of the country, the preferential access to a large number of markets, and the wide endowment of the most important supply in the services sector: human capital. Within this context, the PROSOFT initiative arose as the main strategy to increase IT services and software production and to obtain a competitive advantage in Latin America. Through this initiative, the Mexican government provides the economic resource for certification of IT professionals, software enterprises (organized by clusters), or academics from a previously registered university. The latter has had direct impact on the education of undergraduate students because they began to obtain updated knowledge from their teachers. Another strategy launched by the Mexican government in 2008 is the Program for the Development of the Interactive Media Industry (PROMEDIA) to create the necessary conditions to ensure the growth and consolidations of interactive media industry in Mexico, as well as to increase their international competitiveness through IT. To conclude, in 2009 the initiative "MexicoFIRST" was established with the support of industry, the Secretariat of Economy, and the World Bank with the aim to generate human capital trained in the new generation IT. Through this initiative, 21,395 persons were capacitated and 17,116 were certificated in 2011 (including academics, project managers, programmers, etc.). The resources that this initiative receives come from the World Bank and they are part of the public politic and the PROSOFT programme. Regarding academics training, an agreement with the iCarnegie Institute, from the Carnegie Mellon University, has been signed for training 600 teachers in over 30 universities in Mexico. This staff will be certificated in the methodologies and contents that Carnegie Mellon University uses. Similarly, MexicoFIRST has negotiated preferential prices with Microsoft for the most requested certifications.

3.2 Modern Educational Plan for Organizing People and Activities

Nowadays, there are more than 2,000 universities in Mexico which are coordinated to maintain updated the curricula and establish homogenized programs that promote a better education in computer science and software engineering. In this effort, the National Association of Educational Institutes in Computer Sciences (ANIEI) has highlighted to ensure that the profile of graduates not only achieves the current enterprise requirements but also allows them to explore and participate in new markets and business lines, as well as strengthen the enterprise innovation schemes. Thus, the Secretariat of Economy and the ANIEI proposed the updating of universities' computer science curricula and have created six extra-curricular courses based on an e-learning platform, in order to incorporate the real necessities of the industry into the academic profiles. During 2007, this project was submitted to the e-Mexico committee in order to develop a complete profile of the extra-curricular model. The courses content was placed in a platform, called Capacinet, to give it greater scope, dissemination and availability.

In 2005 the Academia-Industry-Government Society in Information Technologies (IMPULSA) was created. This society has the main objective to constitute an institutional space that allows the articulation of initiatives to promote the use of IT through coordinated efforts of academia, industry and government. During 2007, this society established the project for an Intelligent System of Information in Capabilities of IT Industry (SIICAP) to allow the identification and mapping of the computer sciences graduates in every state of the Mexican republic. Another advance is the signing of a collaboration agreement with the Asia-Pacific Economic Cooperation (APEC) for exchanging academics to update them in teaching techniques. This collaboration agreement has established the basis to launch in Mexico the innovative education and training of undergraduate students through the "APEC Edutainment Park". The results concerning the abilities developed by the human capital would be observed at medium term once students enter the labor market.

3.3 Adaptable Curricula According to Industry Necessities

In 2010, most Mexican universities agreed to establish standardized curricula for undergraduate programs related to Computer Science areas. Currently, Mexican universities' academic councils have defined modular programs with core courses and specialization courses. Thus, a computer science program is defined as a collection of core courses (called 'common basis') which represents a set of subjects related to common programs from the same knowledge area and which have to be taken in the basic stage; and specialization courses which are distributed in three main areas (Artificial Intelligence, Networks and Communications, and Software Engineering) that undergraduate students take in an advanced stage for gaining expertise in one specific area. This elective module program in the Mexican universities comprises basic and advanced courses to better satisfy the students' expectations. Through this module program it is more feasible for undergraduate students to choose an IT domain and can be more effective in helping them to develop more specific skills. However, while developing the modular curriculum, materials are continuously drawn from modern and specialized references (e.g., [2, 6, 17]); its main disadvantage is the limited coverage of the knowledge areas. In Table 2, the included software engineering-related subjects are separated by three categories: the art of practice, process management, and support methods; each one with modules in three levels of complexity: preparatory, transitional, and innovative. These courses were established as an initial remedy to avoid the five malfunctions detected in 2004. For example, a Software Engineering course was enhanced by aggregating specific and new content for those students that want to acquire better technical skills, while for students from a different area of specialization only one course may be sufficient.

Table 2 A modular curriculum for software engineering education

The art of practice	Process management	Support methods
Preparatory level		
Introduction to software engineering	Software quality	Software requirements
Object-oriented software development	Software integration	Development
Component-oriented software development	Software reference models	Business process Modeling
Pattern-based software design		
Software testing		
Software verification and validation		
Transitional level		
Software engineering projects	PSP	Formal specification for software design
Software architecture	TSP	
Secure software	CMMI	COCOMO II
Development	MoProSoft	Six Sigma
Software modeling and analysis		
Innovative level		
Software engineering advanced topics	Project management	Software marketing
Software agile development	Continuous improvement	Risk management for software projects
Software engineering for embedded systems	Software process assessment software metrics	
Software industry	Software factories	

To attempt to reduce the time of student adaptation into industry, courses like Software Industry or Software Factories have been included at the innovative level. To offer a prism view to analyze the real world application domains in different levels, courses covering abstractions such as the notion of objects, patterns, and components were maintained in these curricula, including Object-oriented Software Development, Component-oriented Software Development, and Pattern-based Software Engineering. To raise the awareness of quality, courses such as Software Testing and Software Verification and Validation, Continuous Improvement, Software Metrics, and Software Process Assessments were included in the curriculum. Finally, courses such as Business Process Modeling, COCOMO II, Six Sigma, and Risk Management were included in support of reference models in the process management category at the transitional level. It is important to say that these changes were established in 2010 and some universities still do not have results about their application; in spite of that, the 'old' curricula are reaching the end and updated ones have recently begun: The preliminary results are promising.

4 Evaluation of the Effort to Improve Education

In order to evaluate the results of the abovementioned improvement effort, we have performed a new analysis over the same sample of 210 universities in 2011. As shown in Table 1, we tried to determine the number of offered software engineering-related courses, the number of universities that offer them, and the number of courses that are taught as a single course. Data are shown using the same data source as Table 1. Table 3 summarizes the compared information for 2004 and 2011. We can see clearly that there is a significant improvement measured by the same criteria established in 2004:

- The growth of the total number of offered software engineering-related courses has risen by about 204.10 % from 2004 to 2011, and there is a total of 596 courses since 2011.
- Since 2004, the number of universities that offer software engineering-related courses was up 16 %, from 182 in 2004 to 210 in 2011; and the number of enrollments has increased by 30 % in each software engineering-related course, 17,880 students took these courses in the 2011 academic year.
- The number of universities that offer the modular curricula for software engineering- related courses has risen from 8 in 2004 to 105 in 2011 which accounts for the dramatic decrease of the number of universities that offer a single course from 60 to 27.

To put into perspective the impact of this change, we deeply analyzed some other actions that universities have incorporated into this improvement effort. For example, 12 % of the sample universities have opened a Software Engineering program for undergraduate education. The curricula for these programs are focused on solving four of the malfunctions detected in this study, and universities are working to incorporate strategies and methods for reducing the chasm between courses and the software industry. In this context, some universities have begun to create software factories in support of software engineering education (i.e., [18]) supporting the idea to improve students' abilities through a combination of theory and practice, and focused on industry's necessities. This concept serves as an educational description of how to implement the products that could be produced in a real factory with real clients and demands.

Another important issue, is that universities have begun to recognize the necessity of introducing alternative pedagogical strategies which enable the enhancement of the students' acquired skills. In this sense, in order to improve software engineering education, a general trend exists to emphasize "hands on" experience for the students, related either to industry or a simulated environment. Thus, we found that the most implemented pedagogical strategy is the Project-based Learning (PBL) which is different from traditional learning techniques because the student is introduced to the problem, and he works on the problem and comes up with the solution. Some examples of this change to improve the way to teach software engineering in Mexico are [8, 13, 15, 19]. Furthermore, some other universities are reporting efforts centered

Table 3 Comparison between software engineering-related courses offered in 2004/2011

Course title	2004		2011	
	U	C	U	C
Software engineering	74	82	155	260
Software architecture	6	6	16	18
Software project management	87	87	124	127
Software testing	1	1	12	12
Software integration	0	0	1	1
Component-oriented software development	0	0	1	1
Object-oriented software development	0	0	1	1
Patterns-based software design	0	0	8	8
Software modeling and analysis	67	67	27	27
Software enterprises management	0	0	1	1
Software metrics	1	1	5	5
Software projects development	67	67	5	5
Secure software development	0	0	1	1
Software verification and validation	0	0	3	3
Software engineering advanced topics	0	0	5	5
Business processes modeling	0	0	2	2
Software reference models	0	0	8	8
Software agile development	0	0	3	3
Software projects evaluation	10	10	22	22
Software quality	36	36	64	64
Software process assessment	0	0	1	1
Software product workshop	0	0	2	2
Software industry in the national context	0	0	3	3
Software marketing	0	0	1	1
Professional labor of software engineering	0	0	1	1
Software normativity	0	0	2	2
Formal specification for software design	1	1	3	3
Continuous improvement	0	0	1	1
Software requirements development	1	1	4	4
PSP	0	0	2	3
TSP	0	0	1	1
Six Sigma	0	0	1	1
Risk management for software projects	0	0	1	1
Software factories	0	0	1	1
MoProSoft	0	0	1	1
CMMI	0	0	1	2
Teach a single course	**60**		**27**	
Module program education	**8**		**105**	

U no. of universities, C no. of courses

in other pedagogical strategies such as active learning [21], social design [3], and collaborative learning [1].

Finally, Mexican universities have begun to work with industry to improve the quality of software products through education and certification. In 2006 the SEI (Pittsburgh, USA), Next Process Institute (Kawasaki, Japan) and the Tecnológico de Monterrey (Monterrey, Mexico) signed the "SEI Strategic Partners" agreement with the aim of internationally placing the Mexican software industry through the incorporation of PSP and TSP models [14]. This initiative allowed Mexico to receive knowledge from the SEI about the models as well as authorization to a group of experts and researchers to certify Mexican academics and software enterprises. Currently, Mexico has been positioned as the number one country in the world in relation to the number of people certified in the PSP model, with 61 % of all persons certified in this model around the world. However, this achievement is not enough to consolidate the Mexican industry; it is necessary to ensure that students learn these skills from their certified teachers and that the certified companies use their newly acquired skills in a successful way in the software projects that they develop daily [4, 12].

5 Conclusions and Discussion

It remains a fact that the software industry in Mexico is still very young. However, Mexico is recognizing that this industry is involved in all processes that empower the "new economy" and is paying more attention to the success stories of India—where software is manufactured for internal and international customers—and Brazil—where the government encourages the creation of enterprises in order to effectively compete with United States, Ireland and Canada. It is estimated that in Mexico over 300 software enterprises comprise the software industry and nearly 20 % are formally structured, either as subsidiaries of large international companies, or organizations clearly Mexican, or a combination of both. An important aspect of this fact is that more than 90 % of this sector is integrated by micro-enterprises, most of which present a lack of both structure and a formal approach in their activities for software development, sales and contracts elaboration.

Moreover, these enterprises provide services in isolation with very specific activities to various sectors. Although the approach to the software industry in Mexico has been directed mainly towards the domestic market, it is currently moving towards international markets. However, without skilled human resources, this expansion strategy will not work. In this sense, the role of Mexican universities in the development of the software industry is critical; a greater involvement of curricula with industry is required, and the educational programs must be consistent with current standards of quality and technological development in order to train critical people ready to be productive. The improvement related to students' specialized training in the classrooms would generate, at the undergraduate level, workers skilled in technical programming and software/hardware. Similarly, at the postgraduate level, students will be trained in managerial skills to be better software engineers.

Thus, the analysis presented in this paper is the result of ongoing research that began in 2004 with the aim to establish corrective actions to improve the quality of graduate students in software engineering-related programs. After 7 years of analysis, efforts to improve undergraduate student education have already created a number of new phenomena. Mexican universities are setting up software engineering modules or offering core and specialized software engineering courses and many students are signing up; more universities are opening specialized programs for software engineering at the undergraduate level; and above all, there is an increasingly strong sense of collaboration among teachers and researchers, government, and industry. As a consequence of the 2011 survey, the following promising and encouraging results were observed since 2004:

- The Mexican government has assumed its responsibility in the evolutionary technological project and has established important programs to promote the software industry around the world. Moreover, these programs encourage the universities through government funding and they are focused on training teachers and specialists in specific skills, whom later should deploy the acquired knowledge among their students.
- The Mexican industry has switched its gaze towards academia and has decided to participate more actively in the training of students through collaboration in networks that allow students to get a professional look at their training. In this way, the students not only obtain an academic view of software engineering but they also receive a vision of industry through the tips and guidelines offered by specialists over the network. Other efforts are focused on the establishment of networks to connect the technological necessities of industry with the offer of capabilities and technologies from software engineering researchers of Mexican universities.
- The Mexican academia has reached an agreement in relation to standard curricula for IT-related courses. Since 2004 most curricula are shared in universities and the modular structure is being implemented in software engineering education. The creation of software factories, as an incentive for students' training, has begun to spread among the largest universities to quickly relate students with a real working environment. The 2011 curricula have paid more attention to software process, are more aware in software process quality, show concern to teach how to plan and control software projects, reduce the chasm between courses and the software industry, and avoid the lack of specialization in specific software engineering skills.

Currently, this improvement effort has been recognized as a solid alternative of success and more work is underway. Mexico still has much work to do, but we are accepting our disadvantages and taking into account our strengths to achieve a solid evolution in software engineering education to strengthen the software industry.

Acknowledgments This work is sponsored by ever is Foundation and Universidad Politécnica de Madrid through the Research Chair in Software Process Improvement for Spain and Latin American Region.

References

1. Alanis-Funes, G.J., Neri, L., Noguez, J.: Virtual collaborative space to support active learning. In: Proceedings of the 41th IEEE Frontiers in Education Conference, IEEE Computer Society, pp. 1–6 (2011)
2. Bourque, P., Dupuis, R., Abran, A., Moore, J.W., Tripp, L.: The guide to the software engineering body of knowledge. IEEE Softw. **16**(6), 35–44 (1999)
3. Cárdenas, C.: Social design in multidisciplinary engineering design courses. In: Proceedings of the 39th IEEE Frontiers in Education Conference, IEEE Computer Society, pp. 1–6 (2009)
4. Chen, M.H., Li, T.L.: Construction of a high-performance computing cluster: a curriculum for engineering and science students. Comput. Appl. Eng. Educ. **19**(4), 678–684 (2011)
5. CMMI Product Team: CMMI for systems engineering, software engineering, integrated product and process development, and supplier sourcing (CMMI- SE/SW/IPPD/SS, V1.1). Continuous Representation, CMU/SEI-2002-TR-011, Carnegie Mellon University (2002)
6. CMMI Product Team: CMMI for development, version 1.2, CMU/SEI-2006-TR-008, Carnegie Mellon University (2006)
7. Garcia, I., Pacheco, C., Cruz, D., Calvo-Manzano, J.A.: Implementing the modeling-based approach for supporting the software process assessment in SPI initiatives inside a small software company. In: Lee R. (ed.) Studies in Computational Intelligence, pp. 1–13 (2011)
8. Garcia, I., Pacheco, C.: Using TSPi and PBL to support software engineering education in an upper-level undergraduate course. Comput. Appl. Eng. Educ. Wiley Online Library (2012). doi:10.1002/cae.21566
9. Humphrey, W.S.: PSP: A Self-improvement Process for Software Engineers. Addison-Wesley Professional, Reading (2005)
10. Humphrey, W.S.: TSP Leading a Development Team. Addison-Wesley Professional, Reading (2005)
11. Humphrey, W.S.: Software: the competitive edge. In: International Conference in Software Engineering and Applications, Guadalajara, Jalisco, Mexico (2008)
12. Llorens, A., Llinals-Audet, X., Ras, A., Chiaramonte, L.: The ICT skills gap in Spain: industry expectations versus university preparation. Comput. Appl. Eng. Educ. Wiley Online Library (2010). doi:10.1002/cae.20467
13. Maxinez, D.G., Garcia-Galvan, M.A., Sanchez-Rangel, F.J., Chavez-Cuayahuitl, E., Gonzalez, E.A., Rodriguez-Bautista, R., Ferreyra-Ramirez, A., Aviles-Cruz, C., Siller-Alcala, I.I.: Interactive scenario development. In: Thomas, G., Fleaurant, C., Panagopoulos, T., Chevassus-Lozza, E. (eds.) Recent Researches in Mathematical Methods in Electrical Engineering and Computer Science, pp. 35–39. WSEAS Press, Athens, Greece (2011)
14. Nichols, W.R., Salazar, R.: Deploying TSP on a national scale: an experience report from pilot projects in Mexico, CMU/SEI-2009-TR-011, Carnegie Mellon University (2009)
15. Noguez, J., Espinosa, E.: Improving learning and soft skills using project oriented learning in software engineering courses. In: Proceedings of the 2nd International Workshop on Designing Computational Models of Collaborative Learning Interaction, pp. 83–88 (2004)
16. Oktaba, H.: MoProSoft: a software process model for small enterprises. In: Proceedings of the First International Research Workshop for Process Improvement in Small Settings, Software Engineering Institute, Carnegie Mellon University. Special Report CMU/SEI-2006-SR-001, 2006, pp. 93–101 (2006)
17. O'Regan, G.: Introduction to Software Process Improvement (Undergraduate Topics in Computer Science). Springer, London (2010)
18. Peredo-Valderrama, R., Canales-Cruz, A., Peredo-Valderrama, I.: An approach toward a software factory for the development of educational materials under the paradigm of WBE. Interdisc. J. E-Learn. Learn. Objects **7**: 55–67 (2011)
19. Polanco, R., Calderon, P., Delgado, F.: Effects of a problem-based learning program on engineering students' academic achievements in a Mexican university. Innovations Educ. Teach. Int. **41**(2), 145–155 (2004)

20. Project Management Institute: A guide to the project management body of knowledge (PMBoK guide). Project Management Institute (2004)

21. Ramirez-Hernandez, D., Leon-Rovira, N.: Active learning in engineering: examples at Tecnológico de Monterrey in México. In: de Graff E., Saunders-Smits G.N., Nieweg, M.R. (eds.) Research and Practice of Active Learning in Engineering Education, pp. 56–62. Pallas Publications, Amsterdam, Netherlands (2005)

22. Russell, S.: ISO 9000:2000 and the EFQM excellence model: competition or cooperation? Total Qual. Manag. **11**(4–6), 657–665 (2000)

23. Sampedro, J.L., Vera-Cruz, J.A.: Absorptive capacity of information and knowledge through interfaces in the customized software industry: the case of micro and small-sized Mexican firms. In: Proceedings of the GLOBELICS 6th International Conference, Georgia Institute of Technology, pp. 1–17 (2008)

24. Secretariat of Economy: National plan for development 2007–20012. http://pnd.calderon.presidencia.gob.mx/pdf/PND_2007-2012.pdf (2007)

25. Secretariat of Economy: Current situation of IT in Mexico. National system of IT indicators. http://www.economia.gob.mx/files/transparencia/prosoft_eval_2007.pdf (2007)

26. Secretariat of Economy: Software industry development program. http://www.prosoft.economia.gob.mx/doc/prosoft20.pdf (2008)

Does a Code Review Tool Evolve as the Developer Intended?

Osamu Mizuno and Junwei Liang

Abstract In this study, we intend to assess the improvements of Gerrit. The central concern is "Does Rietveld evolve into Gerrit as the developers intended?" To answer this question, we first compare qualitative features of two code review tools. We then conducted an interview with a developer of Gerrit and obtained the developer's original intention of improvements in Gerrit. By analyzing mined data from code review logs, we try to explain the effects of improvements quantitatively. The result of analysis showed us that the improvements of Gerrit that the developer is expected are not observed explicitly.

Keywords Software evolution · Code review · Open source development · Empirical study

1 Introduction

Mining software repositories has become a current trend of software engineering. Much research using mining technique has been done to discover various issues in software engineering [1]. Repositories mining research often focuses on the quality of software. For example, detection of fault-prone software modules is one of good field of repositories mining [2, 3]. In fact, the number of fault-prone module detection approaches using mining technique has rapidly increases in last 5 years.

In order to assure the quality of software, early detection of defects is recommended. The code review is one of effective ways for such early detection of defects in software [4]. The code review activities include various useful insights for software quality. However, especially in open source software (OSS) developments, records of code review merely remained in a systematic way. Logs of reviews were

O. Mizuno (✉) · J. Liang
Software Engineering Laboratory, Graduate School of Science and Technology,
Kyoto Institute of Technology, Kyoto, Japan
e-mail: o-mizuno@kit.ac.jp

J. Liang
e-mail: j-liang@se.is.kit.ac.jp

© Springer International Publishing Switzerland 2015
R. Lee (ed.), *Software Engineering Research, Management and Applications*,
Studies in Computational Intelligence 578, DOI 10.1007/978-3-319-11265-7_5

mostly on the mailing-list, and researchers needed so many efforts to obtain data from mailing-lists or unstructured data [5].

Recently, for the effectiveness of the code review process in OSS development, tools for code review have been developed. One common weakness of mail based review was the lack of linking between patches and the version control system [6]. The review with tools provides this linkage. For example, a diff made against an older version of a file can be updated by selecting the most reviewed version within a tool, or an approved review can be immediately committed into the version control system.

Not so many open source projects, however, adopt review with tools. Chromium is one of them, which is a successful open source browser project used the code review tool called Rietveld. Rietveld is an open source code review tool based on Mondrian, the internal code review tool used by Google. Both of Rietveld and Mondrian was created by Guido van Rossum who is best known as the author of the Python programming language.

Rietveld was a good tool for code review. The developers, however, appended patches to Rietveld to improve features they want. As a result, they built a new tool for code review from a scratch. The documents of Gerrit says the background as follows [7]:

> Gerrit Code Review started as a simple set of patches to Rietveld, and was originally built to service AOSP. This quickly turned into a fork as we added access control features that Guido van Rossum did not want to see complicating the Rietveld code base. As the functionality and code were starting to become drastically different, a different name was needed. Gerrit calls back to the original namesake of Rietveld, Gerrit Rietveld, a Dutch architect.

For this reason, Gerrit is a successor of Rietveld, and thus it is natural that Gerrit has improved features from Rietveld. Like Rietveld and Gerrit, software is sometimes improved and evolved into another new software. Such improvement is done by various reasons, for example, user's requirements, developer's intuition, performance issues, and so on. Although the effect of improvement should be evaluated, it is hard to measure and thus remained not evaluated.

In this study, we aim to evaluate the improvements of Gerrit. The basic question is "Does Rietveld evolve into Gerrit as the developers intended?" To answer this question, we first compare qualitative features of two code review tools. We then conducted an interview with a developer of Gerrit and obtained the developer's original intention of improvements in Gerrit. By analyzing mined data from code review logs, we try to clarify the effects of improvements quantitatively.

The rest of this paper is organized as follows: Sect. 2 describes the code review process with tools and comparison of code review tools. Section 3 shows the result of interview with a developer of Gerrit and research questions. Mined data from code review repositories are explained in Sect. 4. Section 5 investigates the research questions. The threats to validity are discussed in Sect. 6. Finally, Sect. 7 concludes this study.

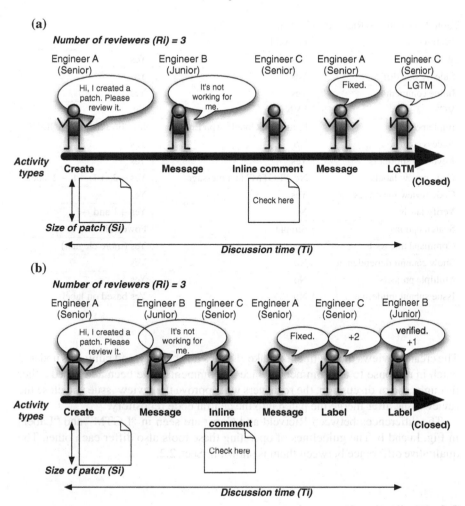

Fig. 1 An example of code review process. **a** Review process with Rietveld. **b** Review process with Gerrit

2 Code Review Tools

2.1 Code Review Process

Figure 1 shows an example process of code review using code review tools. Figure 1a shows a case of Rietveld and Fig. 1b shows a case of Gerrit. In both tools, the fundamental process is similar. First, a developer create a review issue with his/her patch to the source code. Then some developers (reviewers) look at the patch and write messages or inline comments to the creator if they find any problems.

Table 1 Features of Rietveld and Gerrit

Feature	Rietveld	Gerrit
Web-based	Yes	Yes
Side by side diff	Yes	Yes
Inline comments	Yes	Yes
VCS	SVN	Git
Implementation	Python on Google App Engine	Java on J2EE and SQL DB
Access control	No	Yes
Keyboard shortcuts	Yes	Yes
Code review labels	No. "LGTM" in a message	Yes $+2$, $+1$, -1, -2 labels
Code review messages	Yes	Yes
Verify labels	No	Yes $+1$ and -1
Search options	Simple	Powerful
Command line tools	Yes	Yes (more elegant)
Show commit dependencies	No	Yes
Multiple projects	No	Yes
Issue submit guideline	No	Yes based on labels

The creator reviews the comments, make changes if necessary, and publishes updated patch in response to the comments. Once all comments have been addressed either through code or discussion, the reviewers will approve the review issue and close the review issue after merge the patch into the central code repository.

The differences between Rietveld and Gerrit are seen in "LGTM" and "Label" in Fig. 1a and b. The guidelines of operating these tools also differ each other. The qualitative difference between them is shown in Sect. 2.2.

2.2 Qualitative Comparison

Table 1 shows features of Rietveld and Gerrit. Since Gerrit is a successor of Rietveld, most features in Gerrit are enriched from Rietveld.

By comparing features between Rietveld and Gerrit shown in Table 1, the improvement in Gerrit is summarized the following 4 features: (1) Access control, (2) Code review labels, (3) Verify labels/messages, (4) Search options.

The main motivation of Gerrit development is integration with Git. As a result of Git integration, the access control feature is attained to Gerrit.

Shawn Pearce, the main developer of Gerrit, said:

> Access controls in Gerrit are group based. Every user account is a member of one or more groups, and access and privileges are granted to those groups.

Rietveld has no explicit label feature, but has "Quick LGTM (looks good to me)" feature. By pressing the "Quick LGTM" button, the reviewer can quickly post a message with "LGTM". By extending this future, a label system is implemented in Gerrit.

The labels attached on an issue are essential information to determine to accept or decline the code review. Such labels are called as "code review labels" in these code review tools. Rietveld. however, does not have the label, but has the message named "LGTM (Looks Good To Me)". Reviewers attach this message if they find that the code is good for commit. In the development of Chromium, there is not a clear criterion to determine to accept or decline an issue, i.e. if most of participant attached "LGTM", the issue may be accepted.

On the other hand, Gerrit implemented the label system with labels of "+2 (Looks good to me, approved)", "+1 (Looks good to me, but someone else must approve)", "0 (No score)", "−1 (I would prefer that you didn't submit this)", and "−2 (Do not submit)". The +1 and −1 levels are just an opinion where as the +2 and −2 levels are approving or abandoning the review issue. Only review issues having at least one +2 and no −2 labels can be approved. There is no meaning to accumulate these label value. Two +1 labels do not equate to one +2 label [8].

Since Gerrit is a successor of Rietveld, the main purpose of Gerrit is the improvement of code review processes in Google OSS projects. Such improvements are, however, not evaluated quantitatively.

3 Research Questions

3.1 Interview with the Developer

For clarifying the original developer's intentions to develop Gerrit as a successor of Rietveld, we interviewed with Shawn Pearce, who is the main contributor of the development of Gerrit. The questions and answers are as follows:

Q1: What is the motivation to adopt access control of developers in Gerrit? Is it for reducing spammy or trivial commits?

A2: AOSP determined it wanted senior engineers who were conducting code reviews to be able to tick an "approve" box on the web, but not be distracted by the mechanics of downloading a patch, applying it to a local tree, and pushing that to the central server.

To allow anyone to submit a change without doing the manual "download + patch + commit + push" steps from the command line we introduced access controls to determine who can tick the "approve" box, and who can push the "submit" button to deliver the code to the central server automatically.

It enabled a productive boost for the senior engineers that were mostly conducting the code reviews.

Q2: What was the main purpose to adopt code review labels (+2, +1, ...)? We guess that it is for a criterion of patch commits (i.e. if a patch have +2 or larger label, the patch can be committed.) What is the purpose of introducing the verify labels and messages?

A2: Yes. It was a means to realize the access controls to tick the "approve" box. Once we decided we needed an approve box, we built the code-review label.

We then realized we wanted maybe a junior engineer we trust to download the patch, compile it, verify unit tests still pass, etc. and have them mark a different box that says "yup, code compiles!".

This become the verified box. So the code-review OK and the verified OK needed to be done by different people (code-review: senior engineer, verified: junior engineer), so these were different labels.

We then realized it looked bad in the community that nobody else could come along and say "yes I also like this patch". So we expanded code-review to be a range of −2 through 2. Trusted +senior engineers familiar with the project were given access to use the −2 and +2 end of the range. Everyone else (even a random user that stumbles on the site) can use the −1 ⋯ + 1 range to say "yes me too I also like this".

3.2 Research Questions

From the interview with Shawn Pearce, we aim to clarify whether the productive boost for senior engineers is achieved or not. To do so, we state the following research question:

RQ Do code review activities become more productive in Gerrit-based project?

RQ aims to investigate whether the entire review activity in Gerrit-based project is more productive or not.

4 Data Retrieval

4.1 Target Projects

In order to compare the review processes between Rietveld and Gerrit, we need to find review repositories that adopt Rietveld or Gerrit. For Rietveld, we used the Chromium project,[1] a development of Chromium browser by Google and Google Web Toolkit

---

[1] http://www.chromium.org.

Table 2 Format of issue data from Rietveld

Item	Type	Description
i	Nominal	Unique ID for the issue
R_i	Counting	The number of reviewers who appear in the issue i
T_i	Counting	The discussion time for the issue i
S_i	Counting	The size of patch code is the issue i (lines)
N_i^{LGTM}	Counting	The number of "LGTM" messages in code review.

Table 3 Format of issue data from Gerrit

Item	Type	Description
i	Nominal	Unique ID for the issue
R_i	Counting	The number of reviewers who appear in the issue i
T_i	Counting	The discussion time for the issue i
S_i	Counting	The size of patch code is the issue i (lines)
L_i^{max}	Counting	The max value of labels in the issue i.

Table 4 Format of developer commitment data

Item	Type	Description
d	Nominal	Unique ID for the developer
i	Nominal	Unique ID for the issue
A_i^d	Nominal	The action that the developer d takes in the issue i

(GWT) project.[2] For Gerrit, we used the Android project,[3] a development of Android OS projects, and Qt project.[4]

4.2 Data Obtained

We obtained the log of code review activities by our mining tool for code review repositories [9]. We obtained various kind of data from the code review repositories. Tables 2 and 3 show the schema of data related to review issues. Table 4 shows the schema of data related to the developer's activities in review. These tables include only necessary information for later discussions.

---

[2] https://developers.google.com/web-toolkit/.

[3] http://source.android.com.

[4] http://qt-project.org.

5 Do Code Review Activities Become More Productive in Gerrit-based Projects?

5.1 Preliminary Questions

For investigating RQ, we define three preliminary questions as follows:

- How many engineers in review are there in Rietveld-based and Gerrit-based projects?
- How many review issues are there in Rietveld-based and Gerrit-based projects?
- What kind of activities did engineers do in review in Rietveld-based and Gerrit-based projects?

The first challenge is that, sometimes, a developer would like to have more than one accounts to commit to the project, which we called *developer aliases* issue. If we want to answer questions about developer, such as how many engineers involved in project, we should resolve the aliases issue first. We implemented a similar algorithm based on the *levenshtein edit distance* [10] with the approach proposed by Bird et al. [11] to automatically extract developer aliases. We also manually review the result from the web interface of our mining tool, and remove the alias links we thought were questionable.

Table 5 is an overview of data that we obtained for each project. It provide the answers of the first and the second questions. In more detail, the number of reviewers for each issue, R_i, is summarized in Table 6 by projects. The median of reviewers for each issue is one developer for both Rietveld-based projects and Gerrit-based projects. Interestingly, the same result has been also observed in other OSS projects using commit-then-review (CTR) patch review process [6]. Figure 2 shows distribution of R_i by projects.

From Table 5, we can see that the number of issues per unique reviewers varies by projects, not by code review tools. We cannot find significant difference on the number of review issues and the number of unique developers in this table.

Table 5 Overview of data obtained for each project

Project	Chromium (Rietveld)	GWT (Rietveld)	Android (Gerrit)	Qt (Gerrit)
Review issues	82,303	3,294	9,413	34,891
Number of unique reviewers	1,850	295	823	692
Number of issues per unique reviewer	44.49	11.16	11.52	50.42
Number of commitments	444,218	21,719	49,029	359,031
Duration	Sept. 2, 2008–May 24, 2011	Dec. 3, 2008–Aug. 18, 2012	Oct. 21, 2008–Oct. 26, 2012	May 17, 2011–Nov. 28, 2012

Table 6 Descriptive statistics for R_i for each project

Project	R_i (count)			
	Minimum	Median	Mean	Maximum
Chromium	0	1	1.270	19
GWT	0	1	1.249	7
Android	0	1	1.197	18
Qt	0	1	1.407	12

From Table 6 and Fig. 2, we can see the distribution of the number of reviewers for each issue, R_i. By comparing the Rietveld-based and Gerrit-based projects, there are more issues with $R_i = 1$ and less issues with $R_i = 0$ in Rietveld than in Gerrit. By investigating further for R_i, we found that there are many $R_i = 0$ issues in the abandoned issues of Gerrit-based projects, but there is not so many issues with $R_i = 0$ in Rietveld even in abandoned issues. It can be explained that many issues are immediately abandoned after creation in Gerrit.

Figure 3 shows that the ratio of action types for each project. "Create" shows an event that an engineer creates a review issue. "LGTM/label" shows that an engineer gives an "LGTM" message for an issue in Rietveld or a label for an issue in Gerrit. "Message" shows that an engineer gives a message for an issue. Finally, "Inline comment" shows that an engineer gives an inline comment for an issue.

We can see that the ratio of "LGTM/Label" is greater in Gerrit-based projects than Rietveld-based projects, while the "inline comment" activity is smaller in Gerrit-based projects than in Rietveld-based projects. This indicates that developers in Gerrit-based projects prefer to contribute to reviews by voting review or verify labels rather than add inline comments to patch code.

5.2 Main Question

For this research question, we measure how a project is productive by the discussion time for review issues. Generally speaking, if we need less time for an issue to process, we can say it is more productive. We thus translate original research question into as follows:

- Does an issue take less discussion time to review in Gerrit-based projects?

In order to investigate this research question, we define the following metrics:

- T_i: the discussion time for an issue i (h)
- S_i: the size of patch submitted for an issue i (LOC)

In order to consider the difficulty of an issue, we investigate the discussion time normalized by the changed lines of code for an issue, which is expressed as follows:

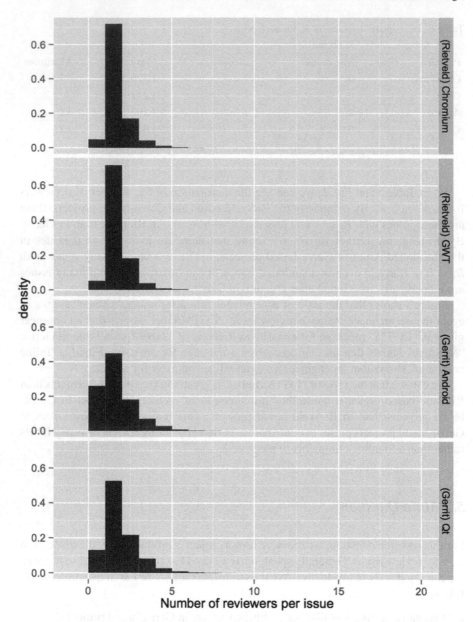

Fig. 2 Density histogram of R_is for all projects

$$\frac{T_i}{S_i}$$

Table 7 shows the descriptive statistics for T_i for each project. The median of T_i in Chromium is 2.199 h and GWT is 15.64 h, while it is 23.04 h in Android and 20.98 h in Qt. It should be noted that the median of T_i in Android is close to 24 h.

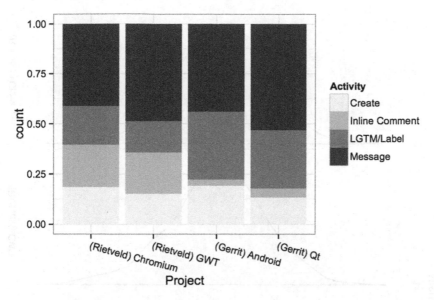

Fig. 3 Histogram of action types for each project

Table 7 Descriptive statistics for T_i for each project

Project	T_i (h)			
	Minimum	Median	Mean	Maximum
Chromium	0.009	2.199	53.73	8746
GWT	0.009	15.640	193.40	8240
Android	0.009	23.040	689.50	8751
Qt	0.009	20.980	136.40	8524

Table 8 Descriptive statistics for T_i/S_i for each project

Project	T_i/S_i (h/line)			
	Minimum	Median	Mean	Maximum
Chromium	0	0.098	4.067	5,211
GWT	0	0.134	12.510	2,325
Android	0	0.874	121.500	8,670
Qt	0	0.934	15.030	5,013

The descriptive statistics for T_i/S_i is shown in Table 8. Table 8 shows that median of T_i/S_i of Rietveld-based projects are significantly smaller than that of Gerrit-based projects.

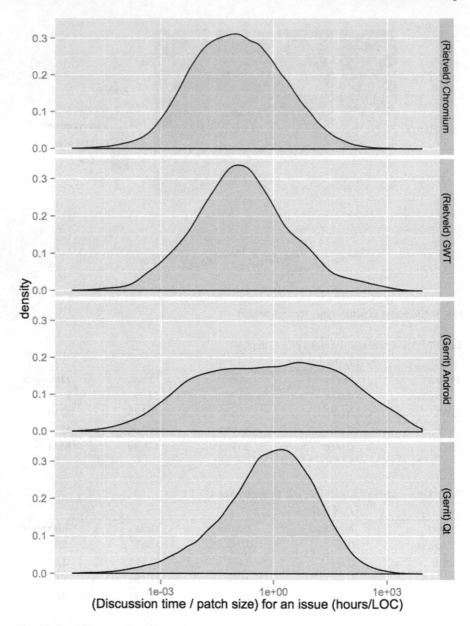

Fig. 4 Density histogram of T_i/S_is for all projects

Figure 4 shows a density histogram for T_i/S_i in all projects. Note that x-axis is shown in log scale. We can see that Gerrit-based projects require more discussion time to review a line of code than Rietveld-based projects.

In Gerrit-based project, it is recommended to pend an issue for 24 h to ensure any interested parties around the world have had a chance to commit. The document of Gerrit says as follows [7]:

> Pending changes are likely to need at least 24 h of time on the Gerrit site anyway in order to ensure any interested parties around the world have had a chance to comment.

On the other hand, Rietveld has no guideline for reviewing time. This could result the discussion time longer in Gerrit-based projects than in Rietveld-based ones.

To conclude, we can say that the answer of this research question is "no". The discussion time for review issues in Gerrit-based projects is longer than that of in Rietveld-based projects.

5.3 Discussion

As Shawn Pearce told us in his interview, there is a notable improvement in Gerrit, the label system. We investigate the effect of the label system from the viewpoint of discussion time. To do so, we state the following question:

- Do LGTM messages in Rietveld and labels in Gerrit affect to discussion time?

Figure 5 shows the distribution of discussion time by the number of LGTM messages, N_i^{LGTM}, for Rietveld-based projects. Figure 6 shows the distribution of discussion time by the max value of labels, L_i^{max}, for Gerrit-based projects. Since LGTMs

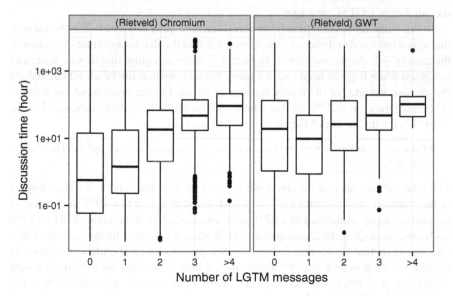

Fig. 5 Discussion time versus the number of LGTM messages for Rietveld-based projects

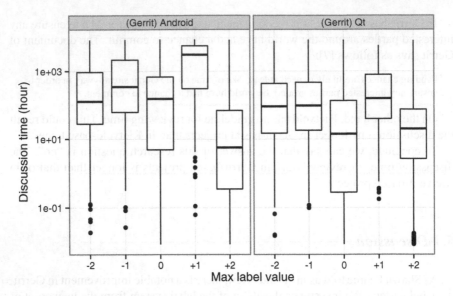

Fig. 6 Discussion time versus the number of labels for Gerrit-based projects

and labels are similar concept, we compare these four projects using LGTM and labels.

In Fig. 5, the discussion time increases according to the increase of LGTM messages. This indicates that if an issue got more LGTM messages, the discussion time becomes longer because there are no clear guideline to finish discussion based on the number of LGTM messages.

In Fig. 6, the discussion time decreases if the max label is +2 or 0. We should note that it is what the developer of Gerrit intended that the discussion time decreases in the case of +2. As we mentioned in Sect. 2.2, there is a guideline that an issue can be closed when it get at least one +2 label. We can say that the label +2 can reduce the discussion time for obviously acceptable issues. On the other hand, we need to investigate the case of $L_i^{max} = 0$ in which the discussion time decreases, too. For the label "0", the document of Gerrit says as follows [12]:

0 No score: Didn't try to perform the code review task, or glanced over it but don't have an informed opinion yet.

The label "0" means that no labels were added by any developers. We then looked at the status of reviews with $L_i^{max} = 0$ and found that 1880 of 1950 such reviews in Android were abandoned and 67 % were abandoned by the creator, 2141 of 2190 such reviews in Qt were abandoned and 76 % were abandoned by the creator. These findings show that developers in Gerrit-based projects often abandoned reviews by themselves even before anyone review that, which made the discussion time very short because no discussion occurred. The similar discussion can be applied to issues with $N_i^{LGTM} = 0$ in Rietveld.

We can conclude that the label system in Gerrit contributes to decrease the discussion time for issues with less problems, while the LGTM messages in Rietveld does not contribute to decrease the discussion time for issues.

6 Threats to Validity

We have several threats to validity in this study. This section discusses on such known threats.

In order to reduce individual variation as far as possible, we tried to find more projects using Rietveld or Gerrit. Unfortunately, we found that there are not so many optional projects for us when we consider project size.

The definition of senior engineers is mainly based on the previous study [13]. There are various definitions of senior engineers in both quantitative and qualitative ways. Since the definition affects the result of analysis directly, we need to refine the definition rigidly.

7 Conclusion

Our findings and contributions are summarized as follows:

- Developers in Gerrit-based projects prefer to contribute to reviews by adding review or verify labels rather than to add inline comments to patch code.
- The discussion time for review issues in Gerrit-based projects is longer than that of in Rietveld-based projects.
- The label system in Gerrit contribute to decrease the discussion time for issues by providing a clear guideline to when an issue can be closed, while the LGTM label in Rietveld does not function to decrease the discussion time for issues.

Acknowledgments The authors would like to express great thanks to Mr. Shawn Pearce, who willingly gave us the answer to our question related to Gerrit. The authors would like to thank Prof. Ahmed E. Hassan and members in Software Analysis and Intelligent Laboratory. This work was supported by JSPS KAKENHI Grant Number 24500038.

References

1. Zimmermann, T., Weißgerber, P., Diehl, S., Zeller, A.: Mining version histories to guide software change. IEEE Trans. Softw. Eng. **31**(6), 429–445 (2005)
2. Catal, C., Diri, B.: Review: a systematic review of software fault prediction studies. Expert Syst. Appl. **36**(4), 7346–7354 (2009). doi:http://dx.doi.org/10.1016/j.eswa.2008.10.027
3. Hata, H.: Fault-prone module prediction using version histories. Ph.D. thesis, Osaka University (2012)

4. Rigby, P.C., Storey, M.A.: Understanding broadcast based peer review on open source software projects. In: Proceedings of 33rd International Conference on Software Engineering, pp. 74–83 (2011)
5. Thomas, S.W.: Mining unstructured software repositories using ir models. Ph.D. thesis, Queen's University (2012)
6. Rigby, P.C.: Understanding open source software peer review: review processes, parameters and statistical models, and underlying behaviours and mechanisms. Ph.D. thesis, BASc. Software Engineering, University of Ottawa (2004)
7. Gerrit code review—system design. URL http://gerrit-documentation.googlecode.com/svn/Documentation/2.5.1/dev-design.html
8. Gerrit code review—a quick introduction. URL http://gerrit-documentation.googlecode.com/svn/Documentation/2.5.1/intro-quick.html
9. Liang, J., Mizuno, O.: Analyzing involvements of reviewers through mining a code review repository. In: Joint Conference of the International Workshop on Software Measurement and the International Conference on Software Process and Product Measurement, pp. 126–132 (2011). doi:http://doi.ieeecomputersociety.org/10.1109/IWSM-MENSURA.2011.33
10. Navarro, G.: A guided tour to approximate string matching. ACM Comput. Surv. (CSUR) **33**(1), 31–88 (2001)
11. Bird, C., Gourley, A., Devanbu, P., Gertz, M., Swaminathan, A.: Mining email social networks. In: Proceedings of the 2006 International Workshop on Mining Software Repositories, pp. 137–143. ACM (2006)
12. Gerrit code review—access control. URL http://gerrit-documentation.googlecode.com/svn/Documentation/2.5.1/access-control.html
13. Baysal, O., Kononenko, O., Holmes, R., Godfrey, M.W.: The secret life of patches: a firefox case study. In: Proceedings of 19th Working Conference on Reverse Engineering, pp. 447–455 (2012)

Analysis of Mouth Shape Deformation Rate for Generation of Japanese Utterance Images Automatically

Tsuyoshi Miyazaki and Toyoshiro Nakashima

Abstract We have been doing research on machine lip-reading. In the process of the study, we proposed a reproduction method of utterance images without voice from Japanese kana. Firstly, sequence of codes that is called "Mouth Shapes Sequence Code" is generated. The Mouth Shapes Sequence Code expresses the order of mouth shapes when a Japanese word is uttered. The utterance images are generated by using the mouth images corresponding to the Mouth Shapes Sequence Code and the deformed mouth shape images generated with morphing. However, the deformation rate of the mouth shapes has been decided from the real utterance images experimentally. Therefore, there were cases in which the utterance images with the sense of incongruity about mouth shape deformation were generated. In this paper, the deformation rate of the mouth shapes is analyzed using real utterance images captured by a high-speed camera, and we propose a generation method of utterance images based on the results. Finally, the mean opinion score of the subjects are shown, and we evaluate the effectiveness of proposed method.

Keywords Lip-reading training · Teaching materials · Computer graphics

1 Introduction

We have been doing research on a technology to support communication of hearing-impaired persons. There are three types of means when hearing-impaired persons communicate with other persons without using voice. They are "sign language",

T. Miyazaki (✉)
Department of Information and Computer Sciences, Kanagawa Institute of Technology,
1030 Shimo-ogino, Kanagawa, Atsugi, Japan
e-mail: miyazaki@ic.kanagawa-it.ac.jp

T. Nakashima
School of Culture-Information Studies, Sugiyama Jogakuen University, 17-3
Hoshigaoka-motomachi, Chikusa, Nagoya, Aichi, Japan
e-mail: nakasima@sugiyama-u.ac.jp

© Springer International Publishing Switzerland 2015
R. Lee (ed.), *Software Engineering Research, Management and Applications*,
Studies in Computational Intelligence 578, DOI 10.1007/978-3-319-11265-7_6

"communication by writing" and "lip-reading". The sign language is effective means, but both of them should have skills of sign language. Therefore, the person to which the hearing-impaired persons can communicate is limited. When the hearing-impaired persons communicate with persons who do not have the skills of sign language, they use the communication by writing. They can communicate with more persons by communication by writing, but it takes time for mutual understanding. Consequently, it is not so effective as the means of conversation. By using the lip-reading, the hearing-impaired persons can communicate with normal listeners the same as communication by writing. In this case, the skills of lip-reading are needed to the hearing-impaired persons, and its acquisition is hard.

When hearing-impaired persons learn the skills of lip-reading, they learn them from the trainer face-to-face or they learn from the video teaching materials of utterance images. Learning effect of the former is higher than the latter, but they cannot learn the skills anytime and the trainer should have the ability of lip-reading. On the other hand, in the latter they can learn the skills at their convenience, but it is difficult to show the movement of lips that uttered any phrase. Therefore, as for the learning using the video teaching materials, the proficiency is apt to vary. If utterance images of any phrases can be generated, we consider that it is available as the teaching materials of lip-reading.

The studies to generate utterance images by using a computer includes "Visual Speech Synthesis". This study generates the images of lips movement synchronized with voice data, namely this study composes the mouth shape and face expression to 3D face model by using the features generated from voice data [1, 2]. In the study on the anthropomorphic spoken dialog agent, a CG model agent moves the mouth in sync with synthesized speech and shows like talking behavior [3]. In these studies, the researchers aim at synchronizing of the lips deformation with the voice. However, we aim at the generation of utterance images that are used as teaching materials to acquire the lip-reading skills. Therefore, the lips movement of the utterance images must be realistic. As a method to decide the mouth deformation, statistical method such as the Hidden Markov Model (HMM) is used [2]. On the other hand, because Japanese has unique relation between voice and mouth shape [4], our study utilizes this association.

We proposed a generation method of utterance images without voice automatically from kana of a Japanese word [6]. It is possible to generate the utterance images of which the mouth shape deforms correctly by using the proposed method. After that, an application software which was carried out on the Android devices was produced experimentally [5]. However, because these images were generated as 30 fps and the deformation rate of mouth shapes was set based on the observation of the authors, the deformation of the mouth shapes was not fluent.

In this study, the movement of lips is analyzed in detail from the images captured by high-speed camera, and the mouth shape deformation rate is derived. The images are generated based on the deformation rate and evaluated by subjects.

Fig. 1 Four blue markers
for lips movement analysis

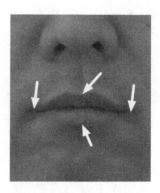

2 Analysis of Mouth Shape Deformation Rate

To analyze the movement of lips when Japanese words and phrases are uttered, the images around lips are captured by a high-speed camera of which the capture rate is 250 fps. To facilitate the analysis, four blue markers that are shown in Fig. 1 are stuck. The area of nostrils is also tracked because the face moves when the person utters.

2.1 Acquire the Sequential Coordinate of the Markers

The flow to detect the markers from a image captured by the camera is shown in Fig. 2. Firstly, the color space is converted into the HSV color space. Secondly, the blue pixels of the markers in the hue plane are extracted. After that, the marker areas (blue pixels) and the background are divided by binarization. However, the areas which are not the markers are misdetected, and there are pixels called holes which are not detected in the marker area either. Because these areas or pixels are small, the misdetected areas are cleared with the Opening process and the holes are filled with the Closing process. Finally, the four marker areas are detected with the process above and the each central coordinate is calculated. The time series coordinates are acquired with the process applied.

For example, the images which were generated with each process are shown in Fig. 3. Figure 3b shows the binarized image that the pixels of Fig. 3a were divided by the marker color. The white pixels belong to the marker and the black pixels belong to the background. However, it still includes misdetected areas and holes. Figure 3c is the image after the Opening process was applied to the Image(b). The misdetected areas are cleaned with this process. Fig. 3d is the image after the Closing process was applied to the image(c). The holes are filled with this process.

Fig. 2 The markers detection
flow

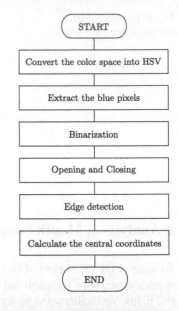

Fig. 3 a Captured mouth
image, **b** Binarized image
with marker color, **c** The
image applied the opening
process and **d** The image
applied the Closing process

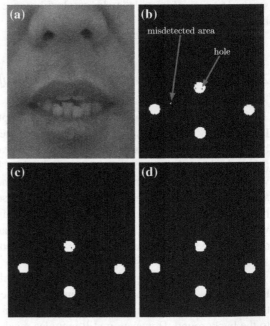

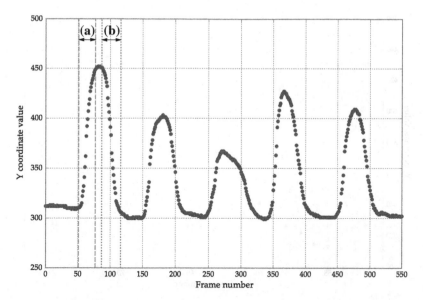

Fig. 4 Sequential Y coordinate values of lower lip

2.2 Approximate Equation of Mouth Shape Deformation Rate

As a result of having analyzed the sequential coordinate values change of four markers, it became clear that upper lip and both corners of the mouth scarcely move. The lower lip also scarcely moved horizontally, on the other hand it very moved vertically. Figure 4 shows the sequential Y coordinate values of lower lip when a person uttered "MA-MI-MU-ME-MO".

In the Fig. 4, the lower lip moves downward in the period changing in upward slant to the right, and it moves upward in the period changing in downward-sloping. Figure 5 shows the values of the period (a) in the Fig. 4, and a curve of cubic equation is also shown. Figure 6 shows the values of the period (b) and the curve of cubic equation. As a result of the Figs. 5 and 6, the movement of lower lip can approximate in a cubic equation.

3 Generation Method of Japanese Utterance Images

The deformed mouth shape images are generated by using the cubic equation. The method that we have proposed previously is used for generating Japanese utterance images.

Firstly, the utterance word is input in Japanese. The word is converted into "Mouth Shape Sequence Code" (hereinafter called MSSC) [4]. The deformed mouth shape

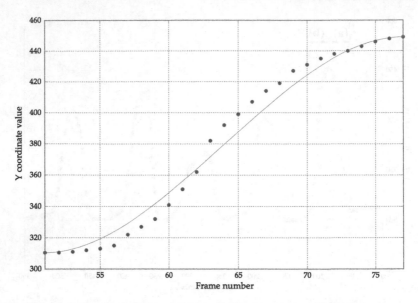

Fig. 5 The coordinate values of the period (a) in the Fig. 4 and a curve of cubic equation

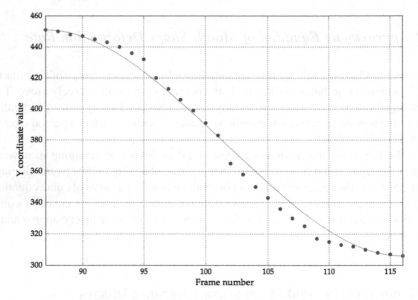

Fig. 6 The coordinate values of the period (b) in the Fig. 4 and a curve of cubic equation

images are generated from the MSSC with the morphing method [6], because the MSSC expresses the sequence of mouth shapes when the word uttered. The images shown in Table 1 are used as the key frames when deformed mouth shape images are generated. However there are lips area, teeth area and buccal area (inside of mouth)

Table 1 Images of basic mouth shape, and the presence of teeth area and buccal area

Basic mouth shape	Image	Teeth area	Buccal area
A		✓	✓
I		✓	✗
U		✓	✗
E		✓	✓
O		✗	✓
X		✗	✗

in the mouth image, and those presence differs from the mouth shapes. Therefore it is necessary to consider this difference when deformed mouth shape images are generated.

For example, when the mouth shape deforms from A to E, each area can be transformed between corresponding areas because the teeth area and the buccal area exist each. On the other hand, when the mouth shape deforms from I to O, the transformation between corresponding areas cannot be processed because the existence of each area is different. Therefor, the deformed area is compounded with gray image.

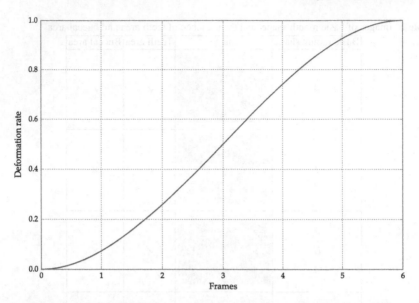

Fig. 7 Deformation rate for the frame

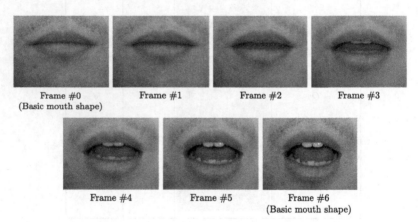

Fig. 8 Generated mouth shape images when closed mouth shape changes into mouth shape /a/

In case the utterance images are displayed in 60 fps, from the result analyzed in Sect. 2.1, the number of the frames required for deformation of the mouth shape is 7. The mouth shape deformation rate for the frame is shown in Fig. 7. As an example, generated mouth shape images are shown in Fig. 8. It shows the mouth shape deformation from the closed mouth to mouth shape /a/. The frames #0 and #6 are the key frame, and the frames from #1 to #5 are generated images.

4 Experiments

We carried out experiments to evaluate the generation method of Japanese utterance images. We evaluated fluency of utterance images and understanding of utterance word to 15 subjects who do not have lip-reading skills. The utterance speed of the images was "Slowly", "Normal" and "Fast". Even if the number of the mouth shape deformation frames is changed by the change of utterance speed, the mouth shape deformation rate can be calculated easily from the approximate equation described in Sect. 2.2.

4.1 Fluency of Utterance Images

We showed the images three times for each utterance speed, and the subject evaluated the images between 1 point to 5 points. The utterance words are shown in Table 2. If the point is high, it means that the mouth shape deformation is fluent. The results of mean opinion score are shown in Table 3, and the chart is also shown in Fig. 9.

As a result, we consider that the images of word #5 (ASESUMENT) are fluent, because the scores are almost 4 for each utterance speed. On the other hand, the images of word #3 (KAMISHIBAI) are not fluent, because the scores are less than 3. We consider that the factor is "MI" and "SHI" of consecutive voice. The both mouth shapes are same because the two voices have same vowel. There were the subjects

Table 2 Utterance words and their mouth shape sequence code for the first experiment

#	Word (in Japanese)	English	MSSC
1.	KATATSUMURI	A snail	-AIA-UXU-I
2.	KAWAKUDARI	Going downstream in a boat	-AUA-UIA-I
3.	KAMISHIBAI	A story told with pictures	-AXI-IXA-I
4.	ASESUMENTO	An assessment	-AIE-UXE-IUO
5.	SUPOTTORAITO	A spotlight	-UXO-UUOIA-IUO

Table 3 Result of mean opinion score of utterance images

#	Word	Normal	Slowly	Fast
1.	KATATSUMURI	3.60	4.20	3.13
2.	KAWAKUDARI	3.93	3.20	3.00
3.	KAMISHIBAI	2.33	2.60	2.53
4.	ASESUMENT	4.07	4.07	3.80
5.	SUPOTTORAITO	3.40	3.47	2.67
Average		3.47	3.51	3.03

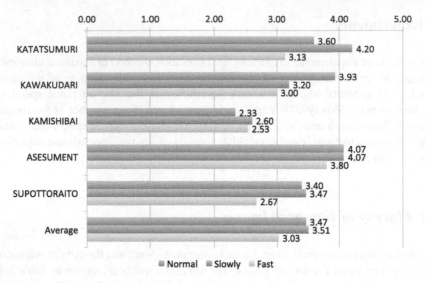

Fig. 9 Bar chart of the result of mean opinion score

who commented that it is difficult to distinguish "MI" and "SHI". It is necessary to improve the display method of mouth when the same vowels continue.

4.2 Understanding of Utterance Word

The utterance images of the word shown in Table 4 were generated. The utterance speed was "Normal" and "Slowly". As well as the former experiment, the utterance images were shown three times to the subjects. However, the subjects were informed about the category hint shown in Table 4 instead of the utterance word.

The understanding rate of each word is shown in Table 5, and the chart is also shown in Fig. 10. The understanding rates of word #2 and #3 were high, but only one

Table 4 Utterance words and their mouth shape sequence code

#	Word (in Japanese)	English	MSSC	Category hint
1.	SUIKA	a watermelon	-U-I-A	Vegetables or fruit
2.	KARASU	a crow	-AIA-U	Bird
3.	FUDEBAKO	a pen case	-UIEXA-O	Stationery
4.	KAWASAKI	Kawasaki city	-AUAIA-I	Name of a city in Japan
5.	UDEDOKEI	a wristwatch	-UIEUO-E-I	Wearable item
6.	SAIBANSYO	a court	IA-IXA-IUO	Landmark

Table 5 Understanding rates of utterance words

#	Word	Normal (%)	Slowly (%)
1.	SUIKA	60.0	73.3
2.	KARASU	80.0	86.7
3.	FUDEBAKO	80.0	86.7
4.	KAWASAKI	66.7	80.0
5.	UDEDOKEI	53.3	73.0
6.	SAIBANSYO	6.7	6.7
Average		57.8	67.8

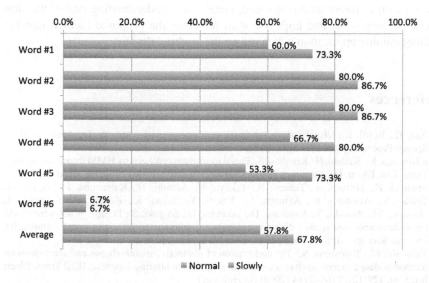

Fig. 10 Bar chart of the correct answer rates

subject understood the word #6. The subjects commented that they could understand the last part of the word #6 but could not understand the former part. In addition, we consider that it was difficult to understand the word #6 because the word had more "Beginning Mouth Shapes" [4] than other words.

However, we consider that the understanding rate was high except for the word #6 because the subjects did not have lip-reading skills. From these results, the evaluation about the fluency of the utterance images was not high in the first experiment, but we consider that the utterance images of which the word can be understood are generated. In addition, in most words, the utterance speed is regulated appropriately, because the understanding rate of slowly utterance is higher than the normal.

5 Conclusion

We proposed the generation method of Japanese utterance images automatically for the teaching materials that the hearing impaired persons acquire lip-reading skills. The deformation of mouth shapes is analyzed by using a camera which can be captured at high frame rate, and the deformation rate of mouth shapes is derived from the results. The experiment of fluency of the utterance images did not show high evaluation. However, we consider that the utterance images showed correct deformation of mouth shape, because the experiment to understand utterance word from the utterance images showed high evaluation. It is effective to generate the images with different utterance speed, because the understanding rate of the slow utterance images became high. We want to reflect the proposed method into lip-reading training application which run on the Android devices.

References

1. Asa, H., Bertil, L.: Visual speech synthesis with concatenative speech. In: Auditory-Visual Speech Processing, pp. 181–184 (1998)
2. Kiyotsugu, K., Satoshi, N., Kiyohiro, S.: Facial movement synthesis by HMM from audio speech. Trans. Inst. Electr. Inf. Commun. Eng. **J83-D-I** I(11), 2498–2506 (2000) (in Japanese)
3. Shinichi, K., Hiroshi, S., Tsuneo, N., Takuya, N., Satoshi, N., Katsunobu, I., Shigeo, M., Tatsuo, Y., Atsuhiko, K., Akinobu, L., Yoichi, Y., Takao, K., Keiichi, T., Keikichi, H., Nobuaki, M., Atsushi, Y., Yasuharu, D., Takehito, U., Shigeki, S.: Design of software toolkit for anthropomorphic spoken dialog agent software with customization-oriented features. Inf. Process. Soc. Jpn. (IPSJ) J **43**(7), 2249–2263 (2002) (in Japanese)
4. Tsuyoshi, M., Toyoshiro, N.: The codification of distinctive mouth shapes and the expression method of data concerning changes in mouth shape when uttering Japanese. IEEJ Trans. Electr. Inf. Syst. **129**(12), 2108–2114 (2009) (in Japanese)
5. Tsuyoshi, M., Toyoshiro, N.: Development of lip-reading training application for smartphones. Multimedia Distrib. Coop Mob. Symp. **2012**, 1863–1868 (2012) (in Japanese)
6. Tsuyoshi, M., Toyoshiro, N., Naohiro, I.: Evaluation for an automatic generation of lips movement images based on mouth shapes sequence code in Japanese pronunciation. In: Proceedings of Japan-Cambodia Joint Symposium on Information Systems and Communication Technology, 2011 (JCAICT 2011), pp. 89–92 (2011)

Predicting Fault-Prone Modules by Word Occurrence in Identifiers

Naoki Kawashima and Osamu Mizuno

Abstract Prediction of fault-prone modules is an important area of software engineering. We assumed that the occurrence of faults is related to the semantics in the source code modules. Semantics in a software module can be extracted from identifiers in the module. We then analyze the relationship between occurrence of "words" in identifiers and the existence of faults. To do so, we first decompose the identifiers into words, and investigate the occurrence of words in a module. Modeling by the random forest technique, we made a model of occurrence of words and existence of faults. We compared the word occurrence model with traditional models using CK metrics and LOC. The result of comparison showed that the occurrence of words is a good prediction measure as well as CK metrics and LOC.

Keywords Fault-prone modules · Semantics · Software engineering · Software code module

1 Introduction

It is said that the faults in software are not evenly distributed to the modules but are concentrated to specific modules [1]. Many studies pointed out that the 20 % of modules include 80 % of faults. We call such fault-injected modules as fault-prone modules. This fact indicates that if we can predict fault-prone modules correctly, the efficiency of testing improves more. For example, Khoshogoftaar said that if we can predict fault-prone module, we can reduce cost on software testing by half [2].

Currently, The author is in Nara Advanced Institute of Science and Technology.

N. Kawashima (✉) · O. Mizuno
Software Engineering Laboratory, Graduate School of Science and Technology,
Kyoto Institute of Technology, Kyoto, Japan
e-mail: n-kawashima@kit.ac.jp

O. Mizuno
e-mail: o-mizuno@kit.ac.jp

© Springer International Publishing Switzerland 2015
R. Lee (ed.), *Software Engineering Research, Management and Applications*,
Studies in Computational Intelligence 578, DOI 10.1007/978-3-319-11265-7_7

As for the measures for fault-prone module detection, various metrics have been proposed so far [1]. For example, CK metrics in Object-oriented code [3, 4], process metrics [5], software structures [6], the metrics from static analysis [7, 8], and other metrics.

Faults are injected in software for various reasons. For predicting faults, it is necessary to make an assumption on the relationship between the cause of faults and the existence of faults. In this study, we made an assumption that the semantics in the software module has an effect to the fault existence. Semantics in a software module can be extracted from identifiers in the module. We then analyze the relationship between occurrence of "words" in identifiers and the existence of faults.

For predicting fault-prone modules, we propose a new method using identifiers of source code. Identifiers are dominant components in the source code and affect the quality of software. Our previous study showed the impact of identifiers to the fault-prone modules prediction [9]. Therefore, we made research questions to analyze effect of identifier to quality of software. The reach questions is described in Sect. 2.2.

In our approach, we first split identifiers to words. Then, we count the number of occurrence of unique words for each file. Using machine learning technique, we made a fault-prone module prediction model with the occurrence of words as instance variables.

The rest of this paper is organized as follows: In Sect. 2, we describe background of this study. In Sect. 3, we describe theory of proposed method. In Sect. 4, we describe a outline of machine learning method used in this study. In Sect. 5, we describe result of experiments. In Sect. 6, we discuss the result of experiments. We conclude this paper in Sect. 7.

2 Background

2.1 Importance of Identifiers

Identifiers are the names of class, variable, function, and so on, in source code of software. It is said that approximately 70 % of the source code of a software system consists identifiers.

A function name by a combination of words that adequately describe the contents of the function can be understood without reading the contents of the code and annotation (comments). When we look at a function call in the code, it is possible to understand what the function does without looking at the body of the function when the function name is adequate. In cases of variables, with a variable name by combining the words that describe the contents of the stored value and type information, developers can easily understand the meaning and role of variable. In this way, identifiers have useful information to understand the contents of source code as well as the comment lines. The understandability and the readability are considered as a part of software quality, and we guessed that they deeply influence the existence of faults in software modules.

Our past research proposed a fault-prone module prediction approach that uses length of identifier and number of word included in a identifier [9]. Yamamoto et al. adopted the number of variable names as predictor variables of fault density prediction models [10]. These approaches used identifiers but did not use words in identifiers.

2.2 Research Questions

In order to clarify the objective of this research, we state the following research questions to be confirmed in this research:

RQ1: Can the count of word included in identifiers of source codes predict the quality of software?
RQ2: Which method is better for predicting fault-prone modules, proposed method or conventional method?

3 Proposed Method

Our objective is to analyze the relationship between the words in identifiers and the occurrence of faults in a module. For this purpose, out proposed method is a predicting model that uses count of words included in identifiers in source code.

3.1 Process

The process of proposed method is shown in the following.
1. We extract identifiers from source code, and split identifiers to words.
2. We count each word in a module.
3. We link the existence of fault in a module and the counted number of words in a module.
4. By a machine leaning technique, we make a fault-prone predicting model using the number of words in a module.

Fig. 1 shows a chart of the process of making fault-prone module prediction model.

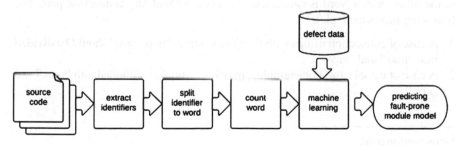

Fig. 1 Chart of making fault-prone module prediction model

3.2 Collecting Data from Projects

For collecting target projects, we used GitHub[1] and PROMISE[2] Data repository.

3.2.1 Defect Data

PROMISE data repository is an open repository of data sets for software engineering. We used data sets in PROMISE repository as target projects, which include defect data, CK metrics, and LOC. Since PROMISE repository does not store source code repositories, we need to obtain source code repositories of corresponding projects from other repository.

3.2.2 Source Code Repositories

GitHub provides the service for software development projects that use the Git revision control system. Various software are developed and maintained in GitHub. We got repositories of target projects from GitHub.

3.2.3 Linking Defect Data to Source Code

The data set from PROMISE is written defect data with corresponding file names and revisions. We linked defect data from PROMISE to source code from GitHub based on the file names and revisions in the data set. We omit the files that cannot be linked with defect data during this process.

3.3 Splitting an Identifier to Words

Here, we explain a rule to split an identifier to words. There are several ways of connecting words in an identifier. Sometimes they are joined by figures or symbols, in the other case, a word is connected to a word without any connection part. The following shows examples:

1. A case of connection using symbols: For example, "max_size" should be divided into "max" and "size".
2. A case of camel case: For example, "maxSize" should be divided into "max" and "size".

---

[1] https://github.com/.

[2] http://promisedata.googlecode.com/.

3. A case of connection by using a figure: For example, "int2char" should be divided into "int", "2", and "char".
4. A case of connecting without any figures or symbols: For example, "maxsize" should be divided into "max" and "size".

In this study, we can handle identifiers in cases of above 1 and 2. For splitting the identifiers in the source code, we used the tool "lscp".[3]

4 Random Forests

We use the random forest for the machine learning algorithm. Since the random forest can handle thousands of input variables, this technique is appropriate to analyze the occurrence of words in source code. When analyzing occurrence of words for identifiers, we need to handle a lot of identifiers in source code. In order to apply the random forest technique, We used a statistical software R.

5 Experiments

5.1 Target Projects

In this experiment, we used Apache Ant and Apache Xalan projects as targets.
Apache Ant
A software tool for automating software build processes. It is implemented using the Java language.
Apache Xalan
A software implements the XSLT 1.0 XML transformation language and the XPath 1.0 language.

These target projects are written in Java language. In this experiment, we use learning data and testing data in a project to see the performance of prediction. For this purpose, we divided data into learning and testing by revisions. Table 1 shows the number of learning files and test files of target projects, and also shows the versions of learning files and test files in the target projects.

Table 1 Training data and testing data of target projects

Project	Training data		Testing data	
	Versions	Number of files	Versions	Number of files
Ant	1.4, 1.5, 1.6	819	1.7	741
Xalan	2.4	682	2.5	762

---

[3] https://github.com/doofuslarge/lscp. lscp is a lightweight source code preprocesser. lscp can be used to isolate and manipulate the linguistic data (i.e., identifier names, comments, and string literals) from source code files.

Table 2 Fault data of target projects

Project	Training data		Testing data	
	Fault injected	Fault free	Fault injected	Fault free
Ant	164	655	166	575
Xalan	110	572	387	375

Table 3 Number of sort of word of project

Project	Number of unique words
Ant	2726
Xalan	4295

Table 2 shows the number of fault injected files and fault free files in the target projects. These data is obtained from the PROMISE data set.

Table 3 shows the number of unique words in each project. We extracted them by the "lcsp" tool.

5.2 Comparison

Since we used the data set published in PROMISE, we can obtain other metrics than the word occurrence. By using these data, we can perform a comparative experiment. The metrics for comparison are CK metrics and LOC, which are both collected in Apache Ant and Xalan projects.

5.2.1 LOC (Line of Code)

The number of lines of code is one of the oldest and most important software metrics. Here, LOC for a class file is measured.

5.2.2 CK Metrics

For the object-oriented design, Chidamber and Kemerer proposed an object-oriented metrics suit. The metrics suit is called "CK metrics". CK metrics suit includes the following 6 metrics:

- WMC (Weighted Methods per Class): The number of methods defined in each class.
- DIT (Depth of Inheritance Tree): The number of ancestors of a class.
- NOC (Number of Children): The number of direct descendants for each class.

Table 4 Confusion matrix of a result of experiment

		Actual	
		Faculty	Not faulty
Prediction	Faulty	True positive (TP)	False positive (FP)
	Not faulty	False negative (FN)	True negative (TN)

- CBO (Coupling Between Object classes): The number of classes to which a given class is coupled.
- RFC (Response for a Class): The number of methods that can be executed in response to a message being received by an object of that class.
- LCOM (Lack of Cohesion of Methods): The number of pairs of member functions without shared instance variables, minus the number of pairs of functions with shared instance variables. If this subtraction is negative, the metric is set to zero.

5.3 Evaluation Measure

Here, we describe evaluation measures of experiments. The result of experiment is shown in the form of Table 4.

- True Positive (TP): True positive shows the number of modules that are classified as fault-prone which are actually faulty.
- False Positive (FP): False positive (FP) shows the number of modules that are classified as fault-prone, but are actually non-faulty.
- True Negative (TN): True negative (TN) shows the number of modules that are classified as non-fault-prone, and are actually non-faulty.
- False Negative (FN): False negative shows the number of modules that are classified as not fault-prone, but are actually faulty.

In order to evaluate the results, we prepare three measures: recall, precision, and accuracy.

5.3.1 Recall

Recall is the ratio of modules correctly classified as fault-prone to the number of entire faulty modules. Recall is defined by Eq. (1).

$$Recall = \frac{TP}{TP + FN} \tag{1}$$

5.3.2 Precision

Precision is the ratio of modules correctly classified as fault-prone to the number of entire modules classified fault-prone. Precision is defined by Eq. (2).

$$Precision = \frac{TP}{TP + FP} \tag{2}$$

5.3.3 F_1 Value

Since recall and precision are in the trade-off, F1-measure is used to combine recall and precision. F1-measure is defined by Eq. (3).

$$F_1 = \frac{2 \times Precision \times Recall}{Precision + Recall} \tag{3}$$

5.4 Accuracy of Prediction

We describe observations on each evaluation measures. This measures evaluate the success of the prediction.

5.4.1 Recall

Table 5 shows that proposed method get high score more than LOC in both project. It also shows that proposed method get high score more than CK metrics in Xalan, but proposed method get high score less than CK metrics in Ant.

Table 5 Evaluate value

Ant			
	Evaluate value		
Method	Recall	Precision	F_1
Proposed method	0.606	0.668	0.636
CK metrics	0.631	0.656	0.643
LOC	0.522	0.650	0.579
Xalan			
	Evaluate value		
Method	Recall	Precision	F_1
Proposed method	0.142	0.625	0.232
CK metrics	0.127	0.700	0.214
LOC	0.127	0.551	0.206

5.4.2 Precision

Table 5 shows that proposed method get high score more than LOC in both project. It also shows that proposed method get high score less than CK metrics in both project.

5.4.3 F_1 Value

Table 5 shows that proposed method get high score more than LOC in both project. It also shows that proposed method get high score more than CK metrics in Xalan, but proposed method get high score less than CK metrics in Ant.

5.5 *Importance of Words*

Figures 2 and 3 show the Gini index of top 10 words in Ant and Xalan, respectively.

Fig. 2 Importance of words (Ant)

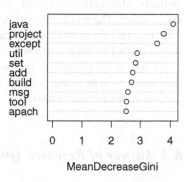

Fig. 3 Importance of words (Xalan)

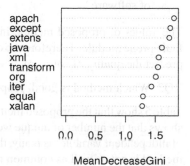

6 Discussion

6.1 Comparison of Number of Independent Variables

The number of independent variables when use CK metrics is 6. The number of independent variables when use LOC is 1. However,the number of independent variables when use proposed method is indefinite. Table 3 shows number of unique words, and it is the number of independent variables. It shows that the number of independent variables is much more than common methods. This means that our approach cannot be handled by a conventional regression analysis due to the number of independent variables.

6.2 Important Words

Figures 2 and 3 show that important word include "java", "except", "apach" in both projects. Among them, "except" is the third highest word in Ant and the second highest in Xalan. The word "except" is a stemmed form and it includes both "exception" and "except". From this fact, we can say that faults frequently appear in the modules that is related to the exception.

This shows a possibility that we can extract bug related words in the context of the source code by using our approach. Analyzing specific words in the source code context is our further research.

6.3 Answer of Research Questions

RQ1 Can the count of word included identifiers of source codes predict the quality of software?

The values of proposed method in Table 5 show that proposed method can predict fault-prone module. Therefore the count of word included identifiers of source codes predict the quality of software.

RQ2 Which method is good, proposed method or conventional method?

Table 5 show that the proposed method is not less than common method. Since Table 3 shows that the number of unique words in a project is numerous. Because the number of independent variables is many, the time of machine learning is long. The proposed method is not as well as common method at this point.

6.4 Threats to Validity

6.4.1 Bug in the Tools for Experiment

If program which used in experiment contains bugs, the result may not be valid. Since we checked the logic of experiment several times, we have a confidence on the program's validity.

6.4.2 Incomplete Data Set

If data set which used in experiment includes incomplete data, the result may be not valid.

6.4.3 Auto Generated Source Code

If source code which objects of experiment was generated automatically, the result may be not valid.

7 Conclusion

We did observations based on result of experiment, we conclude that proposed method can predict fault-prone modules and proposed method is not less than common method in predicting efficient,but proposed method is less than common method in time of machine learning.

For the future issue, if we reveal that tendency of fault-prone words, we know rule of naming identifier which improve quality of source code. In the other, if we choice word which used in independent variable,we can improve the predicting efficient and speed up machine leaning.

Acknowledgments This work was supported by JSPS KAKENHI Grant Number 24500038.

References

1. Hata, H., Mizuno, O., Kikuno, T.: A systematic review of software fault prediction studies and related techniques in the context of repository mining. JSSST Comput. Softw. **29**(1), 106–117 (2012)
2. Khoshgoftaar, T.M., Seliya, N.: Comparative assessment of software quality classification techniques: an empirical study. Empirical Softw. Eng. **9**, 229–257 (2004)
3. Briand, L.C., Melo, W.L., Wust, J.: Assessing the applicability of fault-proneness models across object-oriented software projects. IEEE Trans. Softw. Eng. **28**(7), 706–720 (2002)

4. Gyimóthy, T., Ferenc, R., Siket, I.: Empirical validation of object-oriented metrics on open source software for fault prediction. IEEE Trans. Softw. Eng. **31**(10), 897–910 (2005). http://dx.doi.org/10.1109/TSE.2005.112
5. Ostrand, T., Weyuker, E., Bell, R.: Predicting the location and number of faults in large software systems. IEEE Trans. Softw. Eng. **31**(4), 340–355 (2005)
6. Graves, T.L., Karr, A.F., Marron, J., Siy, H.: Predicting fault incidence using software change history. IEEE Trans. Softw. Eng. **26**(7), 653–661 (2000). http://doi.ieeecomputersociety.org/10.1109/32.859533
7. Nagappan, N., Ball, T.: Static analysis tools as early indicators of pre-release defect density. In: Proceedings of 27th International Conference on Software Engineering, pp. 580–586. ACM, New York, NY, USA (2005). http://doi.acm.org/10.1145/1062455.1062558
8. Zheng, J., Williams, L., Nagappan, N., Snipes, W., Hudepohl, J.P., Vouk, M.A.: On the value of static analysis for fault detection in software. IEEE Trans. Softw. Eng. **32**(4), 240–253 (2006). doi:10.1109/TSE.2006.38. http://dx.doi.org/10.1109/TSE.2006.38
9. Kawamoto, K., Mizuno, O.: Do long identifiers induce faults in software? A repository mining based investigation. In: Proceedings of 22nd International Symposium on Software Reliability Engineering (ISSRE2011), Supplemental Proceedings, pp. 3–1. Hiroshima, Japan, 2011
10. Yamamoto, H.: Software bug density prediction based on variable name (2010)

The Influence of Alias and References Escape on Java Program Analysis

Shengbo Chen, Dashen Sun and Huaikou Miao

Abstract The alias and references escape are often used in Java programs. They bring much convenience to the developers, but, they also give adverse affects on the data flow and control flow of program during program analysis. Therefore, when analyzing Java programs, we must consider the alias and references escape. This paper proposes a static approach to detecting control flow information of programs with alias and references escape. Firstly, it computes the data flow information, including def-use information and alias information caused by references assign and references escape. Secondly, it analyzes the program and gets the control flow information based on the obtained data flow information. Finally, a case study shows that the proposed method can detect control flow information exactly.

Keywords Reference escape · Program analysis · Control flow · Data flow · Alias information

1 Introduction

Different programming languages have their own features. such as variable scope, exception-handling constructs, pointer alias and references escape. These features give a convenient way for the developers to implement the functions of the program. Java is a popular programming language that integrates many useful features.

S. Chen (✉) · D. Sun · H. Miao
School of Computer Engineering and Science, Shanghai University,
Shanghai 200436, China
e-mail: schen@shu.edu.cn

S. Chen
Shanghai Key Laboratory of Computer Software Testing and Evaluating,
Shanghai 201112, China

D. Sun
e-mail: sundashen@shu.edu.cn

H. Miao
e-mail: hkmiao@shu.edu.cn

© Springer International Publishing Switzerland 2015   99
R. Lee (ed.), *Software Engineering Research, Management and Applications*,
Studies in Computational Intelligence 578, DOI 10.1007/978-3-319-11265-7_8

specially, the alias and references escape are wildly used in Java programs when we develop software applications using Java programming language. But usually they may change the control flow of the corresponding programs. As a result, the data flow and control flow information of the program which we get may be incorrect if we do not take the alias and references escape into account when we analyze Java programs.

Program analysis is the process of analyzing the behavior of a computer program. Program analysis has a very widely application range, it provides support for compiler optimization, testing, debugging, verification and many other activities. There are many program analysis techniques, but according to the principle "whether is needed to run the program or not", these techniques could be divided into static analysis techniques and dynamic analysis techniques. Static program analysis is performed without actually executing programs. Generally, the analysis object is source code. Dynamic program analysis is performed by the way of executing the programs, and the target programs must be executed with sufficient test inputs to produce expected outputs. Because of dynamic analysis needs to execute the programs manually, it needs too many test cases and costs too much. Especially for large programs, the disadvantages of dynamic analysis are more obvious. Current analysis techniques are mostly based on static analysis, and in this paper we focus on static program analysis.

Data flow analysis and control flow analysis are the two most widely used static analysis techniques for program analysis. Both of the two methods refer to data flow information and control flow information. Therefore, current popular analysis techniques often combines these two methods. In this paper, we use data flow analysis to get data flow information, and then we use the data flow information to get control flow information. For Java programs, if the influence of alias and references escape is ignored, the data flow information and control flow information we get may be influenced. Therefore, in order to improve the accuracy of data flow analysis, and get the control flow information of the program, we must take the influence of alias information and references escape into account.

In this paper, we propose a approach to get control flow information of Java programs using static programming analysis with large numbers of conditional states branches and alias and reference escape are been considered. The method is based on traditional data flow analysis and it takes alias information and references escape into account. It first constructs the control flow graph of the program, then it computes the def-use information and alias information, and finally it uses the information to analyze the program and get the control flow information. From the results of program analysis, all of the feasible and infeasible program paths are given out.

This paper is organized as follows. Section 2 gives some primary knowledge which will be used in the rest of this paper. The defuse information and alias information are presented in Sect. 3. The case study is given out in Sect. 4. And Sect. 5 states related work. Additionally, some conclusion remarks and future work are given out in Sect. 6.

2 Primary Knowledge

In Java programs, there are two data types: primitive types and reference types. Primitive types define a range of basic data values that can be stored in a variable. The reference types define references to objects of classes, which contain collections of variables and methods that are described by the classes. Data of primitive types can be of either arithmetic or Boolean type. Class and array types are examples of reference types. Any declared variable in a Java program can be of any of the primitive or reference types. In the rest of the paper, we refer to the reference types as usage types.

When reference types are used in a program, alias or references escape may occur. And when it occurs, the data flow information and control flow information of the program may be influenced.

2.1 Reference Escape

If an object is created in a corresponding method, and immediately it is assigned to a non-local variable or a field of non-local variable, or it is passed to another method as the form of parameters or passed out of the internal method as return values, then the lifetime of the object is beyond to the lifetime of its creation environment. The objects which meet the above conditions are called escaped objects.

In this paper, we focus on two different kinds of escape information: (1) A reference escapes if it is returned to other part of the program. (2) A reference escapes if it is passed as a parameter to a method [11]. If an object is created outside the current scope and is accessed via a reference created outside the current scope, the object is already accessible to some part of the current scope. In this case we say that the object has escaped. If an object has not escaped but will be returned via a reference by the method to its caller, we say that the object will escape. Here, we give two classes *A* and *B* as the examples to address our method, as shown in Fig. 1.

In Class *A*, a method *something*() is located in it, and in the method, an object is created and we use a reference *s* point to it. Then the reference *s* is passed into another method *change*() as a parameter in the same class. Then the method *change*() can use the reference *s* to change the state of the object which *s* points to, but other

(a)

```
Class A{
    Public void change(Student a){ ... }
    Public void something( )    {
    Student s = new Student();
    this.change(s);
    ...
    }
}
```

(b)

```
Class B{
    Student doSomething() {
    Student s = new Student();
    return s;
    }
}
```

Fig. 1 Motivating examples of reference escape. **a** Example one, **b** Example two

regions know nothing about this operation except the method *change*(). Therefore, we say that the reference *s* is escaped into the method.

In Class *B*, a method *doSomething*() is located in it. An object is created within the method and it has a reference *s* pointing to it. Different from the first example, it then returns *s* to the method as a return value. For the caller of this method, it can get the reference *s* and change the state of the object which *s* points to. Therefore, we say that the reference *s* will escape out of the definition method as a return value.

2.2 Aliases

In this paper we focus on two different types of aliases in Java program. One type of aliases occurs when a variable with reference types is assigned to another variable, the two references share the same memory location, such as statement "*p = q*". The other is implicit aliases caused by references escape in method calls. In the previous section, we point out that there are two types of references escape: An object escapes if it is returned to other part of the program or passed as a parameter to a method. In this part, we regard both of the two types of references escape as a kind of reference aliases. For example, if a reference *p* is passed as parameter *q* to a method, then *q* is regarded as an alias of *p*. Similarly, in statement "*q = a.change*()", if the method *change*() returned a reference *p* to its caller *q*, then *p* and *q* actually alias.

For both of the two types of aliases, the reference *p* points to the object which the reference *q* points to, and here we can give the definition of aliases in Java programs.

Definition 1 (*aliases*) Two or more references are bound to the same object in some executions of the program.

When aliases occur in a program, and one reference changes the state of the object it holds, the other object holders may be affected by this change in the condition that they do not know what happens, the results caused by this operation may be fatal.

3 Data Flow Information

In order to get the control flow information of the program, first, we construct the control flow graph, and then we use data flow analysis to analyze the data flow of the program, finally we will get the control flow information. At present, there are many programs which have large numbers of branches, and each branch can determine a control flow. Which branch to choose is determined by the value of the conditional statement, and the value of the conditional statement refers to the data flow information. Therefore, in order to get control flow information, we should firstly get the data flow information of the program.

Most approaches of data flow analysis involve decomposing the whole-program analysis into several sub-analyses of individual components, for example blocks, then summarizing the results of these sub-analyses, and memorizing those results for possible later re-use in other calling contexts. Basic block is a basic unit during program execution [2], each basic block has only one entrance and one exit statement, and could only contain no more than one conditional branch statement. The first step of our approach is to compute the def-use information and alias information of each basic block in the control flow graph.

3.1 Def-use Information

Data flow analysis is a common technique for statically analyzing programs and it is used widely for a long time. Traditional data flow analysis include def-use information: (1) reaching definitions, which propagates sets of variable definitions that reach a program point without any intervening writes, and (2) liveness, which determines the set of variables at each program point that have been previously defined and may be used in the future [4].

In Java programs, if the data type of the variable is primitive types, its value could be changed when it is used. But if the data type of the variable is reference types, when it is used, the state of the object which the reference points to may be changed. Here are some sets [6] used in this paper:

- *Def[B]*: the set of variables which is defined or assigned in block *B*;
- *In[B]*: the set of definition variables which could reach block *B*;
- *Gen[B]*: the set of variables which is defined in block *B* and could reach the exit of the block;
- *Kill[B]*: the set of variables which has been defined previously and defined again in block B;
- *Out[B]*: the set of variables which can reach the exit of block *B*.
- *Use[B]*: the set of variables which is used in block *B*.
- *Pred[B]*: the set of blocks that immediately proceed *B* in the control flow graph.

Each block has an associated in and out data flow set, and some other sets which has been listed. The sets including *Def[B]*, *Gen[B]*, *Kill[B]*, *Use[B]* and *Pred[B]*, are easy to compute according to the basic information of block *B*. *In[B]* is the data flow set of definition variables that reach block *B*, it is related to the set *Pred[B]*. *Out[B]* is the set of variables which can reach the exit of block *B*, it has something to do with set *Gen[B]*, *In[B]* and *Kill[B]*. Both of *In[B]* and *Out[B]* can be received by iteratively evaluating the equations until convergence to a solution. The equations [9] are shown as follows:

$$In[B] = \bigcup_{p \in pred[B]} Out[B] \tag{1}$$

$$Out[B] = Gen[B] \bigcup (In[B] - kill[B]) \tag{2}$$

3.2 Alias Information

In Java programs, due to the usage of references type variables, there may be a large number of alias information. In part II, we mentioned that there are two types alias information: one is caused by assigning a reference to another reference, the other is implicit aliases information caused by references escape in method calls. In this part, all alias relations are of the form $<p, t>$, where p and t are references that represent the same object. In most intraprocedural alias analysis, the following sets [13] are used to compute the alias information:

- *Pred(B)*: the block that immediately proceed B in the control flow graph of the program.
- *GEN_Alias(B)*: the set of aliases that generated in block B and can reach the exit of the block.
- *KILL_Alias(B)*: the set of aliases that destroyed in block B.
- *IN_Alias(B)*: the set of aliases that can reach the entry of block B.
- *OUT_Aias(B)*: the set of aliases that can reach the exit of block B.

Here are some explanations about the alias information set:

$$IN\_Alias(B) = OUT\_Alias(p), p \in Pred(B)$$
$$GEN\_Alias(B) = \{\langle p, t \rangle | p \in Alias(p), t \in Alias(t)\}$$

Our method first analyzes all methods which may be called later and summary the alias information at the exit of the method. Then when the method calls occur, the summaries which we have stored before can be re-used to get the alias information without re-analyzing the method. Due to the existence of method calls, here we introduced a new set to gather the alias information caused by method calls.

CALL_Alias(B): the set of aliases which is caused by method calls in block B.

In this paper we focus on the programs which have condition branches, and we use the flowing equation to compute OUT_Alias[B]:

$$OUT\_Alias[B] = (IN[B]\_Alias - KILL\_Alias[B])$$
$$\bigcup GEN\_Alias[B] \bigcup CALL\_Alias[B]) \qquad (3)$$

4 Case Study

We use the Java program shown in Fig. 2 to illustrate our approach. This program is extracted from a large program and it includes alias information and references escape. The alias information is caused by references assign and the two kinds of references escape which we motioned above. The control flow graph of this program is shown in Fig. 3, which we got by means of control flow analysis for the partial Java program as presented in Fig. 2.

Fig. 2 Motivating example
for the program *P*

```
class Example{
        int a;
        public int getA() {return a;}
        public void setA(int a) {this.a = a;}
}
public class Program {
        public static void changeValue(Example e){
C[1]
            e.setA(10);
        }
        public static Example returnExample(){
                Example ee = new Example();
C[2]
                ee.setA(2);
                return ee;
        }
        public static void main(String[] args) {
                int x=1,y=2,z=5,m=3,n;
                Example ex1 = new Example();
B[1]
                Example ex2 = new Example();
                ex1.setA(x);ex2.setA(y);
                n = new java.util.Random().nextInt(10)
                if(n<3){
                        ex1 = ex2;
B[2]
                        ex1.setA(m);
                        if(ex2.getA() == 2){
B[3]
                                z = m-1;
                        }else{
B[4]
                                z = m+1;
                        }
B[5]  }else if(n<7){
                        changeValue(ex1);
B[6]
                        if(ex1.getA() == 1){
B[7]
                                z = m-2;
                        }else{
B[8]
                                z = m+2;
                        }
                }else{
                        ex1 = returnExample();
B[9]
                        if(ex1.getA() == 2){
B[10]
                                z = m+10;
                        }else{
B[11]
                                z = m-10;
                        }
                }
B[12]  System.out.println(z);
        }
}
```

In this paper, we use program paths [7] to describe the control flow information. Program paths here are the basic block sequences of the programs. They include executable paths and infeasible paths. The whole program paths of program *P* are shown as the Table 1.

4.1 Data Flow Analysis

In this part, we use traditional data flow analysis and ignore the alias information and the influence of method calls. The def-use information, including *Out*[*B*] and *Use*[*B*] of each block, is computed by the method which we mentioned in Sect. 3. Then, we can use the def-use information to get the control flow information.

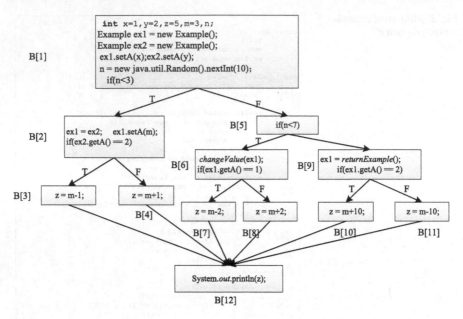

Fig. 3 Control flow graph of the example program *P*

Table 1 All of the program paths

No.	Program path(s)
1	B[1] - B[2] - B[3] - B[12]-exit;
2	B[1] - B[2] - B[4] - B[12]-exit;
3	B[1] - B[5] - B[6] - B[7] - B[12]-exit;
4	B[1] - B[5] - B[6] - B[8] - B[12]-exit;
5	B[1] - B[5] - B[9] - B[10] - B[12]-exit;
6	B[1] - B[5]- B[9] - B[11] - B[12]-exit;

This means that we can get all of the program paths and determine the program paths which are executed or infeasible.

In the control flow graph, the control flow of the program is determined by the value of the predicate expression in the condition branches. In order to get the control flow information, we should use the def-use information to decide the value of predicate expression in each condition branches. First, we analyze the def-use information of variable n in condition branch statement "if$(n < 3)$", after querying the set *Out*[*B*] and *Use*[*B*] of each block, we find that the variable is defined in block *B*[1] and only used in the same block. The value of variable *n* is a random number between 1 and 10. Therefore, the value of the condition branch statement cannot be identified. In this situation, both the truth branch and the false branch of the block *B*[1] can be regarded as feasible to reach. Similarly, for condition branch statement "if$(n < 7)$"

Table 2 The feasibility for each program path

No.	Program path(s)	Feasibility
1	B[1] - B[2] - B[3] - B[12]-exit;	Executable
2	B[1] - B[2] - B[4] - B[12]-exit;	Infeasible
3	B[1] - B[5] - B[6] - B[7] - B[12]-exit;	Executable
4	B[1] - B[5] - B[6] - B[8] - B[12]-exit;	Infeasible
5	B[1] - B[5] - B[9] - B[10] - B[12]-exit;	Infeasible
6	B[1] - B[5] - B[9] - B[11] - B[12]-exit;	Executable

in block $B[5]$, since the value of variable n is a random number between 1 and 10, we regard both the true and false breach as executable.

Then we analyze the condition branch statement "if(ex2.getA() == 2)" in block $B[2]$, the type of the variable $ex2$ is reference types. After searching the sets $Out[B]$ and $Use[B]$ of each block, it is easy to find that the reference $ex2$ is defined in block $B[1]$ and then $ex2$ is used to change the state of the object which it point to in block $B[1]$. The reference $ex2$ is also used in block $B[2]$, but it does not change anything. Therefore, we can conclude that the value of the predicate expression "ex2.getA() == 2" is true, and the truth branch of block $B[2]$ is executable and the false branch can never be reached. For the condition branch statement "if(ex1.getA() == 1)" in block $B[6]$, the variable $ex1$ is also a reference, and it is both defined and used in block $B[1]$. Therefore, the value of the predicate expression "ex1.getA() ==1" is true, the truth branch of the block $B[6]$ is executable and the false branch can never be reached. For the condition branch statement "if(ex1.getA() ==2)" in block $B[9]$, we can use the result of block $B[6]$. After analyzing block $B[6]$, the value of predicate expression "ex1.getA() ==1" is true, so the value of predicate expression "ex1.getA() ==2" is `false`. Therefore, the truth branch of the block $B[9]$ can never be reached and the false branch is executable.

According to the above analysis of the program P, the control flow information of program P can be displayed as Table 2.

4.2 Our Method

In this part, we take the alias information and the influence of method calls into account and we regard $C[1]$ and $C[2]$ as two particular basic blocks. The def-use information, including $Out[B]$ and $Use[B]$ of each block, is computed by the method mentioned in subsection A of Sect. 3, and the alias information, including $CALL\_alias[B]$, $GEN\_alias[B]$, $OUT\_alias[B]$, is computed using the method mentioned in part B of Sect. 3. Then, we can use the def-use information and the alias information to get the control flow information, that is, get all program paths and determine the program paths which are executable or infeasible.

For condition branch statement "if($n < 3$)" in block $B[1]$ and "if($n < 7$)" in block $B[5]$, since there is no alias information in $OUT\_alias[1]$ and $OUT\_alias[5]$, the analysis result of the two blocks is similarly to the result in previous part. That is, the value of predicate expression "$n < 3$" and "$n < 7$" cannot be determined. Therefore, both branches of block $B[1]$ and $B[5]$ are executable.

For condition branch statement "if(ex2.getA() == 2)" in block $B[2]$, after analyzing the alias information and def-use information of the block, we find that it generates an alias pair $<ex1, ex2>$ in this block, the alias pair is in set $OUT\_alias[2]$. Then the reference $ex1$ and the reference $ex2$ are aliases and they point to the same object. Since the reference $ex1$ is in the set $Use[2]$, the statement "ex1.set(m)" changes the state of the object which $ex1$ points to, the reference $ex2$ is influenced by this change. Therefore, the value of predicate expression "ex2.getA() == 2" is false. The false branch of block $B[2]$ is executable and the truth branch will never be reached. For condition branch statement "if(ex1.getA() == 1)" in block $B[6]$, it is easy to find that there is an alias pair $<ex1, e>$ in $CALL\_alias[6]$ and also in $OUT\_alias[6]$. This alias pair is caused by calling the method *chageValue*() and passing the reference $ex1$ into it. Then references $ex1$ and e point to the same object. Since the reference e is in the set $Use[6]$ and it is used in the statement "e.set(10)" in method body $C[1]$, the state of the object which e points to has changed, the variable $ex1$ is influenced by this change. Therefore, the value of predicate expression "ex1.getA() == 1" is false, the false branch of block $B[6]$ is executable and the truth branch could never be reached. For the condition branch statement "if(ex1.getA() == 2)" in block $B[9]$, it is easy to find there is an alias pair $<ex1, ee>$ in $CALL\_alias[9]$ and $OUT\_alias[9]$. This alias pair is caused by calling the method *returnExample*(), after querying the def-use information of the method body $C[1]$, we find that reference ee is defined and used in $C[1]$, the method returns reference ee to its caller $ex1$ at the end of the method. Therefore, it is obvious that the reference $ex1$ does not point to the object which is defined in block $B[1]$. In block $B[9]$, the variable $ex1$ and ee are aliases and they point to the same object. So we can conclude that the value of the predicate expression "ex1.getA() == 2" is true, and the truth branch of block $B[9]$ is executable and the false branch will never be reached.

According to the above analysis of the program P, the control flow information of program P can be achieved as shown in Table 3.

Table 3 The feasibility for each program path using our method

No.	Program path(s)	Feasibility
1	B[1] - B[2] - B[3] - B[12]-exit;	Infeasible
2	B[1] - B[2] - B[4] - B[12]-exit;	Executable
3	B[1] - B[5] - B[6] - B[7] - B[12]-exit;	Infeasible
4	B[1] - B[5] - B[6] - B[8] - B[12]-exit;	Executable
5	B[1] - B[5] - B[9] - B[10] - B[12]-exit;	Executable
6	B[1] - B[5] - B[9] - B[11] - B[12]-exit;	Infeasible

4.3 Analysis

For Table 2, since we analyze the program without considering alias information and the influence of method calls, we get the results that program paths, like No. 1, 3, 6, are executable, and other program paths are infeasible. For Table 3, we take alias information and the influence of method calls into account, we get the results that program paths, like No. 2, 4, 5, are executable, and other program paths are infeasible. After comparing the result of Tables 2 and 3, we can conclude that, since the traditional data flow analysis does not consider alias information and the influence of method calls, it may miss some important data information. As we can see in the Table 3, since we consider alias information and the influence of method calls, the program paths which are executable in traditional data flow analysis may become infeasible, and the program paths which are infeasible in traditional data flow analysis may become executable.

5 Related Work

As an important technology of program analysis, data flow analysis is widely applied into kinds of programs. As for some programming languages have their own features, such as variable scope, exception-handling constructs, pointer alias and references escape. These features give a facility and convenient way for the developers to implement the functions of the program. However, if they are not considered during data flow analysis, the analysis results may be influenced.

The variables may be hidden and covered for their scope in a program. An improved data flow analysis method based on variables scope and traditional dataflow analysis method was proposed to solve the problem of variables being hidden and covered in c program language [6]. For data flow analysis of C++ and Java programs to be correct and precise, the flows induced by exception propagation must be properly analyzed.

Zhang and Jiang et al. [12] adopt a static analysis approach to detecting infeasible paths of programs with exception-handling constructs. In their case study, they only considered the different conditional statements of C++ program, and the pointer alias and reference escape are not taken into account.

As for Java programming language, it also integrates many useful features of modern languages, such as alias, reference escape, exceptions and so on. furthermore, exceptions are also widely used in Java programs which pose new challenge to developers to have data flow analysis. Shelekhov and Kuksenko [10] proposed a method to analyze the exception handling in java programs. The control flow structures for analysis of exception handling are constructed using the information of data flow analysis. However, the alias and reference escape are not considered in their work.

Exception branches, exception plateaus, and exception exits for methods and method calls are introduced as additional control flow structures for analysis of exception handling [5, 9, 10]. Data flow analysis can be classified into two categories: flow-sensitive and flow-insensitive. A flow-sensitive interprocedural pointer alias analysis algorithm and a flow-insensitive interprocedural pointer alias analysis algorithm were presented to improve the efficiency of alias analysis [4, 5]. References escape could only occur in Java programs, the data flow information may be hard to obtain when analyzing a program with references escape. Some new program analysis methods which considering escapes analysis are proposed to get more accurate data flow information [3, 8, 11]. The paper [11] combined pointer and escape analysis for Java programs, and corresponding algorithm was given out. The complete or incomplete Java programs can be analyzed using this method. While in the paper [3], they gave a algorithm for escape analysis of Java objects, and a program abstraction for escape analysis was introduced, using connection graph to build reachability relationships between these objects and references, which can be used to identify the non-escaping objects.

Blanchet [1] used two interdependent analysis, one forward and one backward, for the design and correctness proof of escape analysis for Java. And a method was introduced to prove the correctness of escape analysis.

In this paper, we give out a static approach to detecting control flow information and data flow information of Java programs, and the alias and references escape are considered.

6 Conclusions and Future Work

In this paper, we have proposed an extension approach to the traditional data flow analysis to analyze Java programs. In our method, we take the alias information and references escape into account. We show how implicit alias information occurs when calling a method and the way the alias and references escape influence the control flow information of the program. The analysis of our method is performed in two steps: (1) compute the def-use information and alias information of each basic block in control flow graph. (2) Use the information to analyze the program and get control flow information. The analysis results of the example we give out show that our method can detect the control flow information and demonstrate the influence of alias and references escape on Java program analysis. From the analysis of the alias and reference escape in our program, we can achieve practically all of the feasible and infeasible program path. Consequently, using our proposed method, according to the analysis results of the specific programs, based on the feasible program paths, we can give out the test data. By this means, the test cases can be obtained. So, this method can reduce the quantity of test cases.

However, our approach could only be used in the programs with branches. And we consider the features of Java programs, such as alias and reference escape.

Our future work is to verify the effectiveness of our approach and extend our work to analyze different kinds of Java programs.

Acknowledgments This work was supported by the National Natural Science Foundation of China (NSFC) (61073050, 61170044), Shanghai Leading Academic Discipline Project (J50103). Key Laboratory of Science and Technology Commission of Shanghai Municipality under Grant No. 09DZ2272600.

References

1. Blanchet, B.: Escape analysis for JavaTM: theory and practice. ACM Trans. Program. Lang. Syst. (TOPLAS) **25**(6), 713–775 (2003)
2. Chen, R.: Infeasible path identification and its application in structural test (in Chinese). Doctoral Dissertation, Institute of Computing Technology of Chinese Academy of Sciences, China (2006)
3. Choi, J.D., Gupta, M., Serrano, M., Sreedhar, V.C., Midkiff, S.: Escape analysis for Java. Proceedings of the 14th ACM SIGPLAN Conference on Object-oriented Programming. Systems, Languages, and Applications, OOPSLA'99, pp. 1–19. ACM, NY, USA (1999)
4. Hind, M., Burke, M., Carini, P., Choi, J.D.: Interprocedural pointer alias analysis. ACM Trans. Program. Lang. Syst. (TOPLAS) **21**(4), 848–894 (1999)
5. Jiang, S., Xu, B., Shi, L.: An approach of data-flow analysis based on exception propagation analysis (in Chinese). J. Softw. **18**(4), 832–841 (2007)
6. Jiang, S., Zhao, X.: Data flow analysis based on the variable scope (in Chinese). Comput. Sci. **39**(3), 131–134 (2012)
7. Larus, J.R.: Whole program paths. Proceedings of the ACM SIGPLAN 1999 Conference on Programming Language Design and Implementation. PLDI'99, pp. 259–269. ACM, New York, NY, USA (1999)
8. Lee, K., Fang, X., Midkiff, S.P.: Practical escape analyses: how good are they? Proceedings of the 3rd International Conference on Virtual Execution Environments. VEE'07, pp. 180–190. ACM, New York, NY, USA (2007)
9. Stone, A., Strout, M., Behere, S.: May/Must analysis and the DFAGen data-flow analysis generator. Inf. Softw. Technol. **51**(10), 1440–1453 (2009)
10. Shelekhov, V.I., Kuksenko, S.V.: Data flow analysis of Java programs in the presence of exceptions. Proceedings of the Third International Andrei Ershov Memorial Conference on Perspectives of System Informatics. PSI'99, pp. 389–395. Springer-Verlag, London (2000)
11. Whaley, J., Rinard, M.: Compositional pointer and escape analysis for java programs. Proceedings of the 14th ACM SIGPLAN Conference on Object-oriented Programming. Systems, Languages, and Applications, OOPSLA'99, pp. 187–206. ACM, NY, USA (1999)
12. Zhang, Y., Jiang, S., Wang, Q., Zhao, X.: Infeasible basis paths detection of program with exception-handling constructs. IJACT Int. J. Adv. Comput. Technol. **4**(1), 492–503 (2012)
13. Zhang, Y., Jiang, S., Wang, Q., Zhao, X.: Static approach to detecting infeasible basis paths (in Chinese). J. Front. Comput. Sci. Technol. **6**(2), 144–155 (2012)

Preliminary Evaluation of a Software Security Learning Environment

Atsuo Hazeyama and Masahito Saito

Abstract The importance of software security technologies is increasingly recognized with the increase in services available on the Internet. It is important to foster human resources with knowledge and skills relevant to software security technologies. This paper proposes a learning process for software security and a learning environment that supports the learning process. In the learning process, learners create artifacts for software security (we call them "software security artifacts") from artifacts (we call them "software engineering artifacts") of a traditional software engineering course without dealing with software security, by referring to the knowledge base for software security (standards, methodologies, guidelines, security patterns, and so on). The learning environment supports storage of (1) software engineering artifacts, (2) software security artifacts, (3) a software security knowledge base, (4) rationale and association of the knowledge base with the software security artifacts, and (5) review comments and their association with the software security artifacts. We conducted a preliminary experiment to evaluate the learning process and the learning environment. We confirmed usefulness of the learning process. We also identified some improvements for the knowledge base system and learning environment, such as visualization support and traceability support.

Keywords Software security · Learning environment · Knowledge base for software security · Knowledge sharing

A. Hazeyama (✉)
Department of Information Science, Tokyo Gakugei University,
Koganei-shi, Tokyo 184-8501, Japan
e-mail: hazeyama@u-gakugei.ac.jp

M. Saito
Graduate School of Education, Tokyo Gakugei University,
Koganei-shi, Tokyo 184-8501, Japan

© Springer International Publishing Switzerland 2015
R. Lee (ed.), *Software Engineering Research, Management and Applications*,
Studies in Computational Intelligence 578, DOI 10.1007/978-3-319-11265-7_9

1 Introduction

The number of services available on the Internet is increasing. This trend will continue because of the spread of mobile terminals and/or information electronics in the near future. Along with this trend, computer security becomes increasingly important. In recent years, as increasing number of services has been implemented using software, and with increasing software complexity, the importance not only of network security technologies such as encryption or access control technology, but also of software security technologies has been recognized [12]. Software security deals with security during the whole software development process [12], that is, it is not simply embedded in the network security technologies in a software system but is built into the software system through various types of security activities.

It has been pointed out that it is important to foster human resources with knowledge and skills relevant to software security technologies [12, 13]. We think it is necessary to provide a learning environment for secure software development that learners create artifacts to implement secure software by utilizing body of knowledge on software security. If we can provide such a learning environment, outcomes that are stored in the learning environment will be utilized by other learners as learning materials.

We have been tackling software engineering education in a university [7] and constructed its support system [8]. We focus on web application development based on object-oriented technologies. In our experience, it takes a lot of efforts for novice students to complete even software development that does not include security issues, that is, to elicit functional requirements, and to design, implement, and test software, within one semester. We proposed a software security learning process that used the outcomes of a traditional software engineering course and developed secure software. We also proposed a learning environment for such software security education [9]. This paper describes a preliminary experiment using the learning environment.

The rest of this paper is organized as follows: Sect. 2 discusses the related work. Section 3 proposes the learning process for software security education. Section 4 describes a learning environment that supports the learning process. Section 5 describes some results from the preliminary experiment by using the learning environment. Section 6 describes discussions from the results of the experiment. Finally Sect. 7 summarizes this paper.

2 Related Work

We describe work related to this study from the viewpoint of development support for software security based on knowledge management, and from the viewpoint of software security education.

2.1 Study of Development Support for Software Security Based on Knowledge Management Approach

The SHIELDS project aims at constructing a repository-based secure software engineering environment [6]. The goal of the project is to store and share security models that represent the expertise of experts. The repository model shown corresponds to the knowledge base we describe in this paper.

Barnum and McGraw proposed a knowledge structure for software security whose goal is to form an infrastructure for software security practices [1]. They represented the knowledge structure as a class diagram of seven classes ("Principle," "Guideline," "Rule," "Attack pattern," "Vulnerability," "Exploit," and "Historical risk") and their relationships. We think the knowledge structure proposed by Barnum and McGraw corresponds to the knowledge base we proposed. Our study aims at constructing a learning environment (storing artifacts, association of created artifacts with the knowledge base, and recording the rationale) in which learners can conduct secure software development, referring to the knowledge base.

2.2 Study of Software Security Education

Lester proposed to deal with security in an undergraduate software engineering course according to the increasing importance of software security [11]. Her course dealt with topics on software security ("security life cycle" and "software security in design") twice out of sixteen topics in the course schedule. However, the contents were not clarified.

3 Learning Process

Figure 1 shows an overview of the learning process we proposed [9]. Learners conduct secure software development by inputting artifacts that were created in a software engineering course and were managed by a system [8]. In the software engineering course, various types of artifacts such as requirements specification, use case diagram with use case descriptions, class diagram, database design document, user interface design document, sequence diagram, state chart, source code, and test specification are created. The learning flow, which takes security into consideration after considering the functional requirements of a system, corresponds to a method proposed by Sindre and Opdahl that extracts mis-actors who have hostile intentions, and misuse cases that are hostile actions by mis-actors, and creates misuse case diagrams against normal use cases [17, 18].

However, it is difficult for students who have little professional experience of software security to learn software security without any guidelines and references.

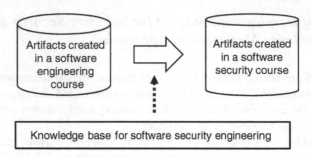

Fig. 1 Learning process for secure software development

In recent years, technologies for software security have become available, for example, methods for security requirement analysis (for example, [16, 19]), security patterns (for example, [15]), standards (for example, Common Criteria [4]), and various types of repositories (for example, CAPEC [3]). We propose using these materials for learning support.

Since we don't suppose specific methods as the learning materials, artifacts created depend on the methods learners select. However, as our software engineering course utilizes the use case diagram, the misuse case diagram will be a promising artifact.

We expect the design rationale to be stored by associating artifacts with the knowledge that was used for their preparation. Quality improvement will be obtained by giving feedback such as reviews and/or inspection to the artifacts.

4 A Learning Environment

We describe a learning environment that supports the learning process. First of all, we discuss requirements the learning environment should provide. Next, we show design of the learning environment, that is, functions of the learning environment and structure of both the whole learning environment and the knowledge base.

4.1 Requirements for a Learning Environment

This section discusses some requirements for a learning environment that supports the learning process we described in the previous section.

1. Requirement 1: The environment should be able to manage artifacts that are inputs for learning about software security and outputs from that learning. Version management of artifacts is mandatory because artifacts created in software development are revised based on review comments and/or change requests.

2. Requirement 2: The environment should allow users to register, revise, and/or delete pieces of knowledge regarding software security (methods, patterns, standards, and so on (hereafter, we call these "knowledge base")). Pieces of the knowledge base may have relationships with each other. Therefore, the learning environment is required to set up relationships and visualize them.

3. Requirement 3: The environment should allow users to register, revise, and/or delete the rationale for artifacts. The environment should enable to establish relationship of the rationale with the artifacts and relationship of the rationale with the knowledge base. Barnum and Sethi emphasized the importance of documentation so that developers could reuse attack patterns that were used in the architecture design in the testing phase [2].

4. Requirement 4: The environment should allow users to give annotations (review comments) to artifacts.

4.2 Design for a Learning Environment

According to the requirements discussed in the previous section, we designed a learning environment. Figure 2 shows the structure of the learning environment. The major functions are as follows.

- Version management of artifacts: The environment manages versions of artifacts so that the revision history of artifacts can be traced (this function is a solution for requirement 1).
- Knowledge base management: It enables to register, revise, browse, delete contents of the knowledge base, and association of pieces of knowledge in the knowledge base (this function is a solution for requirement 2).
- Association of artifacts with the knowledge base and design rationale and cross-referencing using links: This function enables the recording of what learners thought, based on what information, and the relationships between the rationale and the knowledge base (this function is a solution for requirement 3). The environment enables to browse this information in each step of a methodology.
- Annotation of artifacts: The environment enables comments to be linked to artifacts and to be set relationships with the corresponding pieces of knowledge within the knowledge base (this function is a solution for requirement 4).

We explain each function in more detail in the following.

1. Artifact management: The staff enables registration of artifacts created in a software engineering course. The learners enable registration of artifacts they created in a software security course. The revisions of artifacts registered by the learners are managed so that the modification process can be traced. Furthermore, as we deal with several types of software security artifacts, the environment enables them to be related each other (traceability can be preserved between artifacts created in different phases).

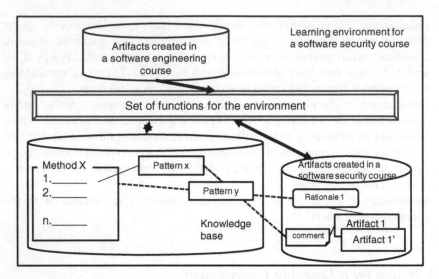

Fig. 2 System structure of the learning environment

2. Construction of the knowledge base: The learning environment enables users to register, revise, browse, and delete contents of the knowledge base. Some pieces of the knowledge base have relationships with each other, and the learning environment enables them to be related to each other.
3. Giving relationships between artifacts and the knowledge base and/or rationale and cross-references: We think it important to record how learners created artifacts for software security with what information in their learning. The environment therefore enables learners to record the rationale and the knowledge base they used to create artifacts, as well as to relate them to each other.
4. Giving comments on artifacts created in a software security course: The environment enables review comments to be given to artifacts created in a software security course and enables relationships between the comments and the artifact to be set. At this time, the environment enables the setting of relationships between the comments and the knowledge base that provides the rationale.

Figure 3 shows a class diagram of the learning environment. Figure 4 shows a class diagram of the knowledge base (please refer the details of the knowledge base to [10]).

We explain the role of each class and the relationships between classes in the following.

"Project" is a class holding the data regarding a project. It has the attributes "project ID," "project name," "project start_date," and "project end_date."

"User" is a class holding the data regarding a user. It has the attributes "user ID," "user name," and "password."

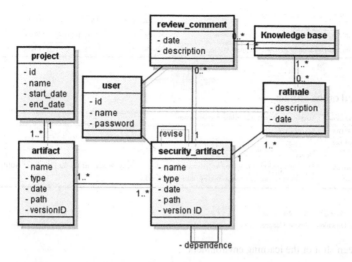

Fig. 3 Class diagram of the learning environment

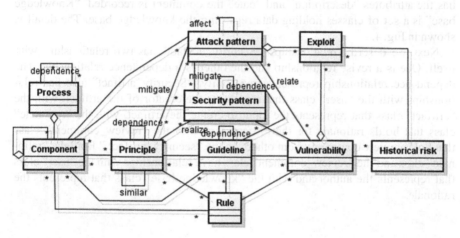

Fig. 4 Class diagram of the knowledge base

"Artifact" is a class holding data regarding artifacts created in a software engineering course. It has the attributes "artifact name," "type of artifact" such as specification, use case diagram, and so on, "created date," "file path," and "version ID."

"Security_artifact" is a class holding the data regarding artifacts created in a software security course. It has the attributes "artifact name," "type of artifact" such as security requirements, misuse case diagram, and so on, "created date," "file path," and "version ID."

"Rationale" is a class holding data regarding the rationale. It has the attributes "description" and "date" the rationale is recorded. "Review_comment" is a class holding review comments on the artifacts created in a software security course. It

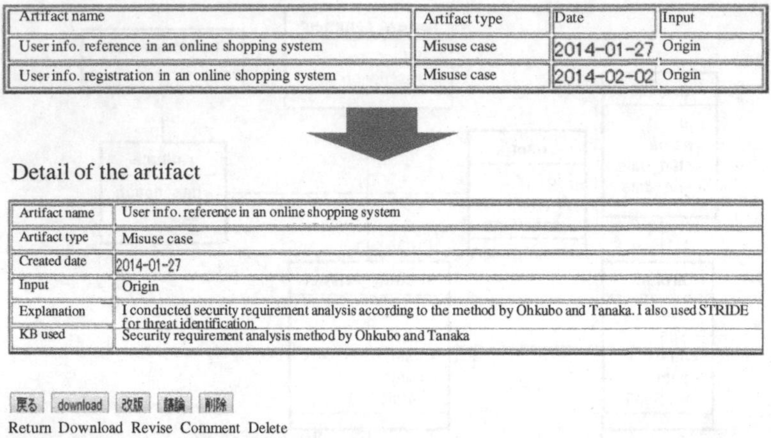

Artifact name	Artifact type	Date	Input
User info. reference in an online shopping system	Misuse case	2014-01-27	Origin
User info. registration in an online shopping system	Misuse case	2014-02-02	Origin

Detail of the artifact

Artifact name	User info. reference in an online shopping system
Artifact type	Misuse case
Created date	2014-01-27
Input	Origin
Explanation	I conducted security requirement analysis according to the method by Ohkubo and Tanaka. I also used STRIDE for threat identification.
KB used	Security requirement analysis method by Ohkubo and Tanaka

戻る download 改版 議論 削除
Return Download Revise Comment Delete

Fig. 5 Screen shot of the learning environment

has the attributes "description" and "date" the comment is recorded. "Knowledge base" is a set of classes holding data regarding the knowledge base. The detail is shown in Fig. 4.

Next, we describe relationships. "Security_artifact" has two relationships with itself. One is a revise relationship and the other is a dependence relationship. The dependence relationship represents traceability. "Security_artifact" also has relationships with the "user" class that represents the creator of the artifact, with the "artifact" class that represents the input to create the artifact, with the "rationale" class that holds rationale for the artifact, and with the "review_comment" class that holds review comments from others on the "security_artifact." Both the "rationale" class and the "review_comment" class have relationships with the "user" class that represents the author and with the "knowledge base" class that represents the rationale.

4.3 Prototype

We implemented a prototype learning environment as a web application.

Figure 5 is a screen shot that displays summary information for an artifact created in a software security course and the detailed information. The artifacts created in a software security course are available for those who have an account of the learning environment in order to show examples to learners to give some clues about how to apply theories or principles to actual secure software development.

5 A Preliminary Experiment

We conducted a preliminary experiment with the prototype learning environment we presented in Sect. 4 in order to evaluate the learning process and the learning environment.

5.1 Overview of the Experiment

We describe an overview of the experiment as follows:

- Participants of the experiment: We applied the learning environment to an advanced software engineering course provided for our graduate school students. Five students participated in the experiment. All of them took the software engineering course and had experience regarding software development by the Project-Based Learning style in their undergraduate program. The advanced software engineering course has provided lectures on software security for 2 months.
- Contents of the knowledge base: This study finally aims to construct a knowledge base that supports full life cycle for secure software development. For the time being, this experiment tries to ascertain potential effectiveness of the learning process and its support learning environment in a qualitative manner. We therefore restricted the scope of development to secure requirement analysis. We stored the following contents into the knowledge base: process for security requirement analysis by Ohkubo and Tanaka [14] (extension to the process by Sindre and Opdahl), guideline for security requirement analysis by Firesmith [5] and representative examples of attack patterns for web applications from the CAPEC dictionary such as SQL injection. We also provided an example project that showed artifacts that correspond to each step of the process [14]. The artifacts were stored as slides in one file.
- Task: The task was given as follows: "analyze security requirements for functions of user information registration and reference of the information in an online shopping system according to the security requirements method by Ohkubo and Tanaka [14] and complete misuse cases. Details of the method and their related information are stored in the knowledge base."

We asked the participants to associate their artifacts with the relevant knowledge as rationale. We also asked them to describe good points and points that need to improve regarding the knowledge base and secure software development with the knowledge base.

5.2 Results

We collected good points and points to be improved for the learning process, the knowledge base and the learning environment. We describe them in the following.

5.2.1 The Learning Process

A student agreed with the learning process that a learner proceeded his/her task by referring to the knowledge base and associated his/her artifacts with the knowledge in the knowledge base. She said it contributed to clarify rationale for secure software development.

5.2.2 The Knowledge Base

We show good points and points to be improved for the knowledge base as follows:

• Good points for the knowledge base

A student said that it was easy for her to search for knowledge because a user selected type of the knowledge she required and that summary and/or references were helpful for her.

Another student said that as many sources regarding software security were originally written in English, this knowledge base written in Japanese was helpful.

• Points to be improved for the knowledge base

Two out of five students pointed out a problem regarding granularity to be analyzed. As Ohkubo and Tanaka provided "enables" relationship, the relationship between abstract level of granularity (for example, STRIDE [20]) and concrete level of granularity (concrete attacks for each category of STRIDE) should be established.

Although a student evaluated selection of the type in knowledge search as good point, another student pointed out a problem that it was difficult to ascertain the whole structure of the knowledge base and a problem regarding lack of awareness on where knowledge exists.

A student pointed out that explanations of knowledge using figures and tables are needed.

5.2.3 The Learning Environment

Some students pointed out insufficient support for setting relationships among different types of artifacts.

6 Discussions

We discuss some enhancements from the results of this experiment.

- Granularity to be analyzed: Some students pointed out the granularity problem to be analyzed. Some guidelines may be required for this problem. In addition, traceability needs to be preserved among different artifacts that present concrete countermeasures for threats (traceability is a requirement for development support environments, not a requirement for the knowledge base).
- Support for making understanding of the structure of the knowledge base easy: In the current version, the contents of the knowledge base are represented in a tabular form. However, as Fig. 4 shows, the meta-model of the knowledge base has complicated relationships. Actual knowledge is instances of each class. Therefore, it has a very complicated network structure. Visualization support is needed to understand such a complicated network structure.
- Enrichment of the contents: A student pointed out that explanations of knowledge using figures and tables were needed. We provided an example project for the security requirement analysis process stored in the knowledge base, however, the linkage from the knowledge base to the example project is not supported (converse linkage was supported). This item is also enhancement one.
- Several types of dependence relationships exist among artifacts created in software development. In our study, not only dependence relationships between artifacts created in different phases in a project but also those between artifacts created in different steps within a method exist. We found traceability was quite important in this experiment. Treatment from abstract level of knowledge to concrete level of knowledge should be surely traced. This is an important enhancement item the learning environment should be implemented.

In this experiment, most students created their artifacts according to the example project stored in the learning environment. A student created his artifacts different from other students. All students gave presentation regarding their artifacts and discussed them one another in this experiment. Our learning process supposed to review. However, review is a kind of defects detection process. From the viewpoint of learning, we found students learned from other learners' artifacts. Positive learning by associating artifacts with the knowledge base for effective usage of the knowledge base is expected effects of this learning environment.

7 Conclusions

This paper has proposed a learning process and a learning environment for software security. The environment stores contents for software security such as methodology, processes, patterns, guidelines, and so on in the knowledge base, and learners proceed to learn about software security by referring to the available information.

Furthermore, recording their rationale promotes reflection by learners. We have also presented some results from a preliminary experiment by using the learning environment.

We have to enhance the knowledge base management system and the learning environment according to the findings we identified in the experiment, especially visualization support and traceability support. We will also enrich the contents of the knowledge base. Hereafter we will conduct an experiment that deals with full life cycle of secure software development.

Acknowledgments This study was partially supported by the Grant-in Aid for No. (C) 22500910 and No. (C) 26330394 from the Ministry of Education, Science, Sports and Culture of Japan.

References

1. Barnum, S., McGraw, G.: Knowledge for software security. IEEE Secur. Priv. **2**, 74–78 (2005)
2. Barnum, S. Sethi, A.: Attack patterns as a knowledge resource for building secure software. http://capec.mitre.org/documents/Attack_Patterns-Knowing_Your_Enemies_in_Order_to_Defeat_Them-Paper.pdf (2014). Accessed 10 June 2014
3. CAPEC: http://capec.mitre.org (2014). Accessed 10 June 2014
4. Common Criteria: http://www.commoncriteriaportal.org/ (2014). Accessed 10 June 2014
5. Firesmith, D.G.: Engineering security requirements. J. Object Technol. **2**(1), 53–68 (2003)
6. Hakon, P., Ardi, M.S., Jensen, J., Rios, E., Sanchez, T., Shahmehri, N., Tondel, I.A.: An architectural foundation for security model sharing and reuse. In: Proceedings of the International Conference on Availability, Reliability and Security 2009, pp. 823–828 (2009)
7. Hazeyama, A.: A case study of undergraduate group-based software engineering project course for real world application. In: Proceedings of the First International Symposium on Tangible Software Engineering Education (STANS2009), pp. 39–44 (2009)
8. Hazeyama, A., Kobayashi, Y.: Collaborative software engineering environment centered around artifacts management and communication support. In: Workshop on Software Engineering Symposium 2008 (SES2008), pp. 5–6 (2008) (in Japanese)
9. Hazeyama, A. Shimizu, H.: Development of a software security learning environment. In: Proceedings of the 13th ACIS International Conference on Software Engineering, Artificial Intelligence, Networking and Parallel & Distributed Computing (SNPD2012), pp. 518–523 (2012)
10. Hazeyama, A.: Survey on body of knowledge regarding software security. In: Proceedings of the 13th ACIS International Conference on Software Engineering, Artificial Intelligence, Networking and Parallel/Distributed Computing (SNPD2012), pp. 536–541 (2012)
11. Lester, C.Y.: A practical application of software security in an undergraduate software engineering course. Int. J. Comput. Sci. Issues **7**(3), 1–10 (2010)
12. McGraw, G.: Software security. IEEE Secur. Priv. **2**(2), 80–83 (2004)
13. Ohkubo, T.: Effectiveness of security analysis technologies in corporations. IPSJ Mag. **50**(3), 230–234 (2009) (in Japanese)
14. Ohkubo, T., Tanaka, H.: A proposal of an efficient security requirements analysis method. J. IPSJ **50**(10), 2484–2499 (2009) (in Japanese)
15. Schumacher, M., Fernandez-Buglioni, M., Hybertson, D., Buschmann, F., Sommerlad, P.: Security Patterns: Integrating Security and Systems Engineering. Wiley, New York (2006)
16. Shimizu, H., Hazeyama, A.: A proposal of security requirements elicitation method by misuse cases in web application development. In: Proceedings of the 73th Annual Conference of IPSJ (2011) (in Japanese)

17. Sindre, G., Opdahl, A.L.: Eliciting security requirements by misuse case. In: Proceedings of the 37th International Conference on Technology of Object-Oriented Languages and Systems (TOOLS-Pacific 2000), pp. 120–131 (2000)
18. Sindre, G., Opdahl, A.L.: Eliciting security requirements with misuse cases. Requirements Eng. J. **10**, 34–44 Springer (2005)
19. SQUARE: http://www.cert.org/sse/square/square-pubs.html (2014). Accessed 10 June 2014
20. STRIDE: http://msdn.microsoft.com/ja-jp/magazine/cc163519.aspx (2014). Accessed 10 June 2014

Expression Strength for the Emotional Scene Detection from Lifelog Videos

Atsushi Morikuni, Hiroki Nomiya and Teruhisa Hochin

Abstract For the purpose of retrieving emotional scenes from a lifelog video database, and showing them, we propose a criterion to measure the strength of emotions. Conventionally, the retrieval of the emotional scenes has been performed by distinguishing the kind of the emotion of a person. However, precisely judging the importance of the scene is difficult without the consideration of the strength of the emotion. Therefore, we introduce a criterion called expression strength in order to measure the strength of emotions based on the amount of the change of several facial feature values. The effectiveness of the expression strength for the detection of the emotional scenes with smiles is shown through an experiment using a lifelog video data set.

Keywords Lifelog · Video retrieval · Facial expression recognition · Emotion

1 Introduction

Due to the recent improvement of the performance of video cameras and storage devices, anyone can easily make a large quantity of multimedia data (e.g., the growth records of children). For such a reason, *lifelog* has attracted attention [1, 2]. Lifelog is the recordings of everyday life and can be recorded as various types of data such as texts, images, and videos. Particularly, video data can be made easily and contain a wide variety of useful information. Therefore, we focus on lifelog videos in this study.

As recording videos becomes easy, an enormous amount of video data can be stored in the video databases. This makes the retrieval of lifelog videos quite difficult.

A. Morikuni (✉) · H. Nomiya · T. Hochin
Department of Information Science, Kyoto Institute of Technology, Goshokaido-cho,
Matsugasaki, Sakyo-ku, Kyoto 606-8585, Japan
e-mail: m2622043@edu.kit.ac.jp

H. Nomiya
e-mail: nomiya@kit.ac.jp

T. Hochin
e-mail: hochin@kit.ac.jp

© Springer International Publishing Switzerland 2015
R. Lee (ed.), *Software Engineering Research, Management and Applications*,
Studies in Computational Intelligence 578, DOI 10.1007/978-3-319-11265-7_10

Consequently, a considerable amount of valuable lifelog data is not utilized. Hence, an efficient and accurate retrieval of useful video scenes is indispensable to fully make use of lifelog videos.

For the better utilization of the lifelog videos, an effective retrieval method has been proposed [3]. This method detects impressive scenes from a lifelog video by using the facial expressions of a person in the video. It can retrieve emotional scenes that a person expresses a kind of facial expression. However, the intensity of the emotion (i.e., the intensity of the facial expression) cannot be estimated. For example, it can detect "smile" but it cannot distinguish "giggle" from "laughter."

In order to improve the retrieval quality of emotional scenes, we introduce *expression strength*, which is the criterion to measure the strength of facial expression. The expression strength is calculated using several salient points on a face called *facial feature points*.

The effectiveness of the expression strength is evaluated by using a lifelog video data set and an image data set widely used for the evaluation of the quality of facial expression recognition. As a result, we reveal that the expression strength is able to estimate the intensity of facial expression and to find the scenes that the users want to retrieve.

The remainder of this paper is organized as follows. Section 2 presents some related works. Section 3 describes the facial features and the computation of the expression strength using the facial features. Section 4 shows the experiment to evaluate the usefulness of the expression strength. Finally, Sect. 5 concludes this study.

2 Related Works

2.1 Video Retrieval

Several video-sharing sites (e.g., YouTube [4]) are widely used to retrieve videos from huge video databases. Many of them provide tag-based retrieval methods. In order to use the retrieval system, some users post their videos with several tags (or keywords) describing the contents of the videos and others search the videos by means of the tags given to the posted videos. This kind of retrieval system will be quite useful for publicly accessible videos.

On the other hand, the owner of lifelog videos must give the tags to each video one by one since most of lifelog videos are private ones. Despite the burden of the tagging tasks, there are only a few scenes that the user wants to watch them again [5]. Hence, the tag-based retrieval is not suitable for lifelog video retrieval.

Content-based video retrieval methods provide more intuitive search frameworks [6]. They can retrieve various scenes by means of such as key frames (key images) on the basis of the similarity of some kinds of feature values obtained from the videos and the key frames. They do not require tagging tasks but it is difficult to retrieve useful scenes because selecting appropriate key frames is not easy.

2.2 Emotional Scene Detection Based on Facial Expression

In general, the users want to retrieve an interesting or impressive scene from lifelog videos. The person in such a scene will express a certain emotion (e.g., surprise, happiness, etc.). Therefore, finding the emotional scenes will be helpful to retrieve useful scenes. Because the emotion can be estimated from facial expressions, an emotional scene detection based on facial expression recognition has been proposed [3].

This method detects emotional scenes by means of the recognition of facial expressions in each frame image of a video. For each frame image, several facial features are computed from the positional relationships of a few facial feature points. Then, all the frame images in a video are classified into one of the predefined facial expressions. Finally, the emotional scenes are detected according to the classification result.

This method can retrieve emotional scenes without troublesome tasks such as tagging. However, it cannot estimate the intensity of the emotion. Estimating the intensity of emotion will be important because a scene with stronger emotion will be more interesting and impressive.

We thus introduce a criterion called *expression strength* for the estimation of the intensity of the emotion. Additionally, we conduct an experiment to estimate the strength of the emotions and verify the hypothesis that the users want to retrieve the scenes with strong emotions.

3 Expression Strength

The expression strength is a criterion to measure the strength of the emotion on the basis of the movement of salient points on a face. We call the points *facial feature points*. The movement of the facial feature points is represented by some facial features defined as the positional relationships of the facial feature points.

3.1 Facial Feature Points

As shown in Fig. 1, the set of facial feature points consist of 42 points on the following components of a face:

- Left and right eyebrows: 10 points ($p_1, \ldots, p_{10}$)
- Left and right eyes: 18 points ($p_{11}, \ldots, p_{28}$)
- A mouth: 14 points ($p_{29}, \ldots, p_{42}$).

These facial feature points are obtained by using an application software called FaceSDK 4.0 [7].

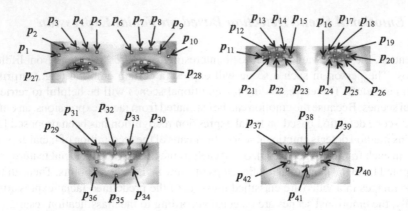

Fig. 1 Facial feature points

3.2 Facial Features

We define the following 11 types of facial features to estimate the expression strength.
These facial features are based on the positional relationships of the facial feature
points and are considered to be associated with the change in facial expressions.

1. Gradient of right and left eyebrows: f_1

This feature value is based on the gradients of the right and left eyebrows (denoted
by a_r and a_l, respectively) computed by using least squares. The gradients of the
right and left eyebrows are obtained from the facial feature points $\{p_1, \ldots, p_5\}$ and
$\{p_6, \ldots, p_{10}\}$, respectively. This feature value is defined as Eq. (1).

$$f_1 = \frac{a_l - a_r}{2} \tag{1}$$

2. Distance between eyes and eyebrows: f_2

Using the average distance between the facial feature points on the eyebrows and
the corresponding facial feature points on the upper side of eyes, the value of this
feature is obtained through Eq. (2).

$$f_2 = \frac{\sum_{i=1}^{10} ||\mathbf{p}_i - \mathbf{p}_{i+10}||}{10 \cdot L} \tag{2}$$

Here, L is a normalization factor for the difference of the size of a face. It is defined as
the distance between the center points of left and right eyes, that is, $L = ||\mathbf{p}_{27} - \mathbf{p}_{28}||$.

3. Area between eyebrows: f_3

This feature value is given by Eq. (3) as the area of the quadrangle formed by connecting four facial feature points $p_5, p_6, p_{16},$ and p_{15} located at the inner corners of eyebrows and eyes.

$$f_3 = \frac{S(p_5, p_6, p_{16}, p_{15})}{L^2} \tag{3}$$

Here, $S(p_{P_1}, \ldots, p_{P_m})$ is the area of a polygon formed by connecting m facial feature points $p_{P_1}, \ldots, p_{P_m}$.

4. Area of eyes: f_4

This facial feature is the normalized areas of two octagons formed by the facial feature points on the circumference of the left and right eyes, defined by Eq. (4).

$$f_4 = \frac{1}{2L^2} \{S(p_{11}, p_{12}, p_{13}, p_{14}, p_{15}, p_{23}, p_{22}, p_{21}) \tag{4}$$
$$+ S(p_{16}, p_{17}, p_{18}, p_{19}, p_{20}, p_{26}, p_{25}, p_{24})\}$$

5. Vertical-to-horizontal ratio of eyes: f_5

Based on the ratio of the distance between the top and bottom points to the distance between the left and right points on the left and right eyes, this feature value is defined by Eq. (5).

$$f_5 = \frac{1}{2} \left(\tan^{-1} \frac{\|\mathbf{p}_{22} - \mathbf{p}_{13}\|}{\|\mathbf{p}_{15} - \mathbf{p}_{11}\|} + \tan^{-1} \frac{\|\mathbf{p}_{25} - \mathbf{p}_{18}\|}{\|\mathbf{p}_{20} - \mathbf{p}_{16}\|} \right) \tag{5}$$

6. Area of the circumference of a mouth: f_6

This feature value is defined by Eq. (6) as the normalized area of the octagon formed by connecting eight facial feature points located on the circumference of a mouth.

$$f_6 = \frac{S(p_{29}, p_{31}, p_{32}, p_{33}, p_{30}, p_{34}, p_{35}, p_{36})}{L^2} \tag{6}$$

7. Area of inner circumference of a mouth: f_7

Similar to the sixth feature value, this feature value is defined by Eq. (7) as the normalized area of the octagon formed by connecting eight facial feature points located on the inner circumference of a mouth.

$$f_7 = \frac{S(p_{29}, p_{37}, p_{38}, p_{39}, p_{30}, p_{40}, p_{41}, p_{42})}{L^2} \tag{7}$$

The sixth feature value are influenced by the thickness of the lips which can vary depending on the type and the intensity of the facial expression. On the other hand, this feature value is hardly affected by the thickness of the lips.

8. Vertical-to-horizontal ratio of the circumference of a mouth: f_8

Based on the ratio of the distance between the top and bottom points to the distance between the left and right points on the circumference of a mouth, this feature value is defined by Eq. (8).

$$f_8 = \tan^{-1} \frac{||\mathbf{p}_{35} - \mathbf{p}_{32}||}{||\mathbf{p}_{30} - \mathbf{p}_{29}||} \tag{8}$$

9. Vertical-to-horizontal ratio of the inner circumference of a mouth: f_9

Similar to the eighth feature value, this feature value is defined by Eq. (9) based on the ratio of the distance between the top and bottom points to the distance between the left and right points on the inner circumference of a mouth. This feature value is also insensitive to the thickness of the lips.

$$f_9 = \tan^{-1} \frac{||\mathbf{p}_{41} - \mathbf{p}_{38}||}{||\mathbf{p}_{30} - \mathbf{p}_{29}||} \tag{9}$$

10. Vertical position of the corner of a mouth: f_{10}

This feature value represents how high the position of the corner of a mouth is. It is defined by Eq. (10).

$$f_{10} = \frac{(y(p_{29}) + y(p_{30})) - (y(p_{32}) + y(p_{35}))}{|y(p_{32}) + y(p_{35})|} \tag{10}$$

Here, $y(p)$ is the y-coordinate of the facial feature point p. If the mean value of the y-coordinate of the facial feature points on the corner of a mouth is larger than that of the facial feature points on the top and bottom of a mouth, f_{10} becomes positive. Thus, a larger value of f_{10} represents a higher vertical position of the corner of a mouth.

11. Angles of corners of a mouth: f_{11}

This feature value is the average value of the angles of the left and right corners of a mouth. The angle of the left (right) corner is formed by connecting the three facial feature points located on the left (right) corner of a mouth. It is given by Eq. (11).

$$f_{11} = \frac{A(p_{29}, p_{31}, p_{36}) + A(p_{30}, p_{33}, p_{34})}{2} \tag{11}$$

where A is the function to compute the angle formed by three facial feature points p, q, and r. A is defined by Eq. (12).

$$A(p, q, r) = \cos^{-1} \frac{(\mathbf{p} - \mathbf{q}) \cdot (\mathbf{p} - \mathbf{r})}{\|\mathbf{p} - \mathbf{q}\| \, \|\mathbf{p} - \mathbf{r}\|} \tag{12}$$

For each frame in a video, the above feature values are computed and the feature vector $(f_{i1}, \ldots, f_{i11})$ is obtained. Here, the j-th feature value obtained from the i-th frame from the beginning of the video is denoted by f_{ij}.

3.3 Expression Strength

The expression strength is defined for a single frame taking into consideration of the tendency that the feature value is proportional to the strength of emotion. In order to accurately estimate the strength of emotion, the baseline of feature value is first determined using a training data set. The training data set consists of several emotional and nonemotional frames. An emotional frame is the frame that a person in the frame image expresses a certain emotion. On the other hand, a nonemotional frame is the frame that the facial expression of a person in the frame is neutral. The training set is manually prepared prior to the computation of the expression strength.

The training data set T is represented as $T = \{g_1^e, \ldots, g_{N_e}^e, g_1^n, \ldots, g_{N_n}^n\}$. Here, g_i^e and g_i^n are the i-th emotional and nonemotional frames, respectively. N_e and N_n are the numbers of emotional and nonemotional frames in the training set, respectively. N_e and N_n have to be determined experimentally.

Because of the personal difference of the facial expressions, it is quite difficult to estimate the expression strength directly from the feature values. Hence, the baseline of feature values are computed to diminish the personal difference of the feature values.

The baseline feature value is determined for each facial feature. The baseline of the j-th facial feature S_j ($j = 1, \ldots, 11$) is defined by Eq. (13).

$$S_j = \frac{\sum\limits_{i=1}^{N_e} f_{eij} + \sum\limits_{i=1}^{N_n} f_{nij}}{N_e + N_n} \tag{13}$$

Here, f_{eij} (f_{nij}, respectively) is the j-th feature value of the i-th emotional (nonemotional) frame in T.

The expression strength is computed on the basis of the difference between the feature values obtained from a frame and the baseline feature values. The expression strength of the i-th frame in a video E_i is defined by Eq. (14).

$$E_i = \sum_{j=1}^{11} (f_{ij} - S_j) \tag{14}$$

Therefore, the higher value of the expression strength represents the stronger emotion.

4 Experiment

4.1 Hypotheses

We make some hypotheses in regard to the usefulness of the expression strength for emotional scene retrieval. Through this experiment, we verify the following hypotheses:

1. The expression strength is able to represent the intensity of facial expressions.
2. The expression strength becomes high in the scenes that the users want to retrieve.
3. The expression strength is not affected by the movement of a mouth unrelated to the expression of emotions, which mainly appears in conversations.
4. The expression strength can be used for various persons.

Note that we focus on the emotion of happiness, which is one of the six basic emotions [8], because most of users will want to retrieve the scenes with happiness. The emotion of happiness often leads to the facial expression of smiles. Hence, we estimate the expression strength of smiles in this experiment.

4.2 Experimental Settings

4.2.1 Lifelog Videos

The experiment was conducted by eight subjects (termed Subjects A, B, C, D, E, F, G, and H), all of which were 21–28 year-old male university students. The subjects were divided into two disjoint groups (denoted by α and β) consisting of four subjects. Subjects A, B, C, and D belong to α and the other subjects belong to β.

For each subjects, the scenes of playing cards were recorded four times by using a web camera placed in front of the subject in order to record the subject's face. The i-th video of Subject X is denoted by X_i ($i = 1, 2, 3, 4$). For example, A_1 means the first video of Subject A. The specification of the web camera is shown in Table 1.

Apart from the web cameras, a video camera was used to record the scenes of playing cards including all the subjects in a group at the same time. Thus, a total of eight lifelog videos were recorded by the video camera. The i-th video for α is

Table 1 Specifications of video cameras

Specifications	Web camera	Video camera
Model	ELECOM UCAM-DLU130HWH	Sony HDR-CX560V
Frame size (in pixels)	640 × 480	1,440 × 1,080
Frame rate	30fps	30fps
Video format	WMV	MPEG4 AVC/H.264

denoted by α_i ($i = 1, 2, 3, 4$). Similarly, the i-th video for β is represented by β_i. The specification of the video camera is shown in Table 1.

The videos recorded by the web cameras were used to estimate the expression strength of the subjects. Because of the high frame rate of these cameras, we selected frames from each video after every 10 frames in order to reduce the computational cost. For example, a total of 1,800 frames are used for a 10 min video.

On the other hand, the videos recorded by the video camera were used to show the videos to the subjects. The subjects were asked to select the scenes they wanted to retrieve from the two videos recorded by the video camera. Subjects A, B, C, and D were asked to select the scenes from α_3 and α_4. The other subjects were asked to select the scenes from β_3 and β_4. The scene was specified by the beginning time and the ending one in seconds. The length of each video is about 10 min.

4.2.2 Learning Settings

Both of the parameters N_e and N_n in Eq. (13) were experimentally set to 12, taking into consideration of the tradeoff between the accuracy and efficiency of the estimation of the expression strength.

Hence, the training set for a certain video consists of 12 emotional and 12 non-emotional frame images. They were randomly selected from all the frame images of the video from which FaceSDK accurately detected the facial feature points. It was determined by one of the authors whether the feature points were accurately detected or not. All the frame images other than the training images were used as the test set.

4.2.3 Strength of Smile

In order to evaluate the estimation accuracy of the expression strength, we first classify the smile into three levels according to the strength of the smile. We call the three levels of smile *Smile1*, *Smile2*, and *Smile3*. The accuracy of the expression strength is evaluated by the average values of the estimated expression strength for the frames of *Smile1*, *Smile2*, and *Smile3*.

The reason for this classification is that the correct expression strength must manually be determined to evaluate the estimation accuracy of the expression strength, but it is very difficult to manually determine the correct expression strength in real number. It seems to be reasonable that classifying the smile into three levels because many people will be able to discriminate the strength of a smile in only a few levels.

The definitions of *Smile1*, *Smile2*, *Smile3*, and the other facial expressions are shown in Table 2. These are defined by one of the authors. He classified all the frames in the videos recorded by the web cameras into one of the expressions described in Table 2.

Table 2 The definition of facial expressions

Facial expression	Definition
Smile1	The subject pulls up the corners of his mouth and slightly opens his mouth
Smile2	The subject lowers the corners of his eyes and smiles with his mouth open
Smile3	The subject laughs with his mouth wide open
Expressionless	The subject shows no expression
Others	The subject shows a certain expression other than smile

4.3 Experimental Result

4.3.1 Verification of the First Hypothesis

We computed the expression strength for the videos A_1, B_2, C_2, D_2, E_1, F_1, G_1, and H_1 because most of the facial feature points were correctly detected in these videos. The average expression strength for each facial expression is shown in Table 3. Note that the frames other than *Smile1*, *Smile2*, *Smile3*, and *Expressionless* are excluded and that Subjects E and G have no frame of *Smile3*.

For most of the subjects, the average expression strength for each facial expression has the relationship that "*Expressionless* < *Smile1* < *Smile2* < *Smile3*." This result indicates that the expression strength can appropriately represent the intensity of smile.

The expression strength of *Smile1* is less than that of *Expressionless* in regard to Subject F. The subject's facial expression of *Smile1* is very similar to that of *Expressionless*, compared with the other subjects. Although the accuracy of the estimation of the expression strength can slightly be affected by the personal difference, we can verify the first hypothesis to some extent from this experimental result.

Table 3 Average expression strength for each facial expression

Video	Expressionless	Smile1	Smile2	Smile3
A_1	−0.163	0.136	0.166	0.258
B_2	−0.345	−0.070	0.055	0.129
C_2	−0.170	0.108	0.111	0.129
D_2	−0.165	0.109	0.201	0.405
E_1	−0.035	0.098	0.383	N/A
F_1	−0.284	−0.361	0.187	0.364
G_1	0.043	0.276	0.312	N/A
H_1	−0.339	0.195	0.242	0.328

Table 4 Average expression strength for the scenes selected by each subject

Video	Sel.	Nonsel.	Expressionless	Smile1	Smile2	Smile3
A_4	0.001	−0.129	−0.179	0.170	0.195	0.241
B_4	−0.033	−0.251	−0.405	−0.007	0.095	0.135
C_4	−0.006	−0.094	−0.210	0.023	0.131	0.148
E_4	0.231	0.003	−0.035	0.098	0.223	0.268
F_4	0.012	−0.025	−0.042	−0.002	0.107	0.214
G_4	0.210	0.067	−0.153	0.153	0.312	0.367
H_4	0.025	−0.222	−0.039	0.195	0.242	0.328

4.3.2 Verification of the Second Hypothesis

We compared the estimated expression strength for each facial expression with that in the scenes selected by the subjects. The videos used in this experiment are A_4, B_4, C_4, E_4, F_4, G_4, and H_4 because the facial feature points are accurately detected in these videos. The video of Subject D was not used because the facial feature points were not accurately detected from both D_3 and D_4 (Note that the subjects selected the scenes only from α_3, α_4, β_3, and β_4).

The average expression strength is shown in Table 4. The columns "Sel." and "Nonsel." represent the average expression strength obtained from the frames selected by the subjects and the other frames, respectively.

For all the subjects, the average expression strength of the selected scenes is higher than that of the other scenes. There are a total of 22 scenes selected by all the subjects. 20 scenes out of the 22 scenes include the frames of *Smile2* and/or *Smile3*. From the fact that the expression strength for *Smile2* and *Smile3* is considerably high for all the subjects, the second hypothesis can be verified.

4.3.3 Verification of the Third Hypothesis

Since some of the facial features are associated with the facial feature points on a mouth, the conversations between the subjects may affect the expression strength. In order to clarify the influence of the conversations on the expression strength, we computed the expression strength of the scene of conversations using the same videos as used in the first experiment.

The average expression strength is shown in Table 5. The columns "Conv." and "Nonconv." represent the expression strength of the conversation scenes and non-conversation scenes, respectively. Note that both scenes do not include any of *Smile1*, *Smile2*, and *Smile3*.

The average expression strength of the conversation scene is much less than that of *Smile1* for most of subjects. Although the average expression strength of the conversation is higher than that of *Smile1* for Subject F, it is much lower compared with

Table 5 Average expression strength for the scenes of conversations

Video	Conv.	Nonconv.	Expressionless	Smile1	Smile2	Smile3
A_1	−0.163	−0.214	−0.039	0.136	0.166	0.258
B_2	−0.345	−0.468	−0.177	−0.070	0.055	0.129
C_2	−0.170	−0.358	−0.134	0.108	0.111	0.129
D_2	−0.165	−0.356	−0.028	0.109	0.201	0.405
E_1	−0.035	−0.120	0.050	0.098	0.383	N/A
F_1	−0.284	−0.385	−0.164	−0.361	0.187	0.364
G_1	0.043	−0.013	0.042	0.276	0.312	N/A
H_1	−0.339	−0.645	−0.136	0.195	0.242	0.328

the expression strength of *Smile2*. From this experimental result, the conversation has very little influence on the expression strength and the third hypothesis can be verified.

4.3.4 Verification of the Fourth Hypothesis

For the verification of the effectiveness of the expression strength for various persons, we used the Cohn-Kanade AU-Coded Facial Expression Database [9], which is widely used for the evaluation of the facial expression techniques. This data set contains the sequences of facial images of 18–30 year-old males and females. We used the image sequences of 18 subjects in this experiment. The image sequence starts with a neutral face and ends with a full smile (i.e., the first image is a neutral face and the last image shows a full smile) as shown in Fig. 2. Hence, the intensity of the smile is proportional to the image number. The number of images in an image sequence depends on the subject and varies from 12 to 33. The average number of images in an image sequence is about 20.9.

Fig. 2 Example of the image sequence in Cohn-Kanade data set

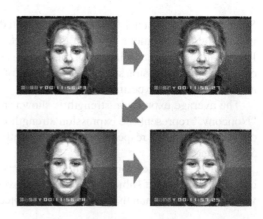

Fig. 3 Expression strength
for Cohn-Kanade data set

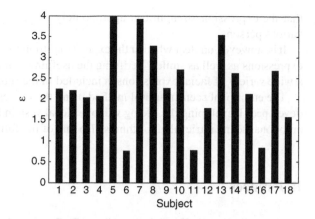

Because the number of images in an image sequence is small, it is impossible to obtain sufficient training images from this data set. Hence, in this experiment, we define the expression strength for a video (denoted by ε) while the expression strength is defined for a frame in the above experiments. The value of ε is defined by Eq. (15).

$$\varepsilon = \sum_{i=1}^{m} \left(\sum_{j=1}^{11} f_{(M-i)j} - \sum_{j=1}^{11} f_{ij} \right) \tag{15}$$

Here, M is the number of images in the image sequence. This value is defined on the basis of the difference of the feature values between the first m images and the last m images in the image sequence. We set the value of the parameter m to 3 considering the average length of the image sequence. The expression strength (i.e., the value of ε) of each subject is shown in Fig. 3.

For all the subjects, the expression strength is positive. From this result, it can be detected by the expression strength that the intensity of the facial expression in the images of a smiling face is higher than that in the images of a neutral face. Although there are personal differences of the expression strength, smiles can be detected from all the subjects by means of the expression strength. Therefore, the fourth hypothesis can be verified from this experimental result.

5 Conclusion

For the purpose of an accurate retrieval of the emotional scenes in lifelog videos, we propose a criterion called expression strength in order to measure the intensity of the emotion. The expression strength is efficiently computed by the geometric computation in regard to the facial feature points. The experimental result showed that the expression strength can appropriately estimate the strength of smiles for various persons.

It is, however, unclear whether the expression strength is effective for diverse facial expressions as well as smiles. Verifying the usefulness of the expression strength for a wide variety of facial expressions is included in the future work.

The emotional scene retrieval method using the expression strength have not yet developed. Developing the lifelog video retrieval system by introducing an efficient emotional scene retrieval algorithm is also one of the future works.

References

1. Aizawa, K., Hori, T., Kawasaki, S., Ishikawa, T.: Capture and efficient retrieval of life log. In: Proceedings of Pervasive 2004 Workshop on Memory and Sharing Experiences, pp. 15–20. (2004)
2. Gemmell, J., Bell, G., Luederand, R., Drucker, S., Wong, C.: MyLifeBits: fulfilling the Memex vision. In: Proceedings of the 10th ACM International Conference on Multimedia, pp. 235–238. (2002)
3. Nomiya, H., Morikuni, A., Hochin, T.: Emotional video scene detection from lifelog videos using facial feature selection. In: Proceedings of 4th International Conference on Applied Human Factors and Ergonomics, pp. 8500–8509. (2012)
4. YouTube: http://www.youtube.com/ (2014). Accessed 1 Apr 2014
5. Shimura, S., Hirano, Y., Kajita, S., Mase, K.: Experiment of recalling emotions in wearable experience recordings. In: Proceedings of the 3rd International Conference on Pervasive Computing, pp. 19–22. (2005)
6. Hu, W., Xie, N., Li, L., Zeng, X., Maybank, S.: A survey on visual content-based video indexing and retrieval. IEEE Trans. Syst. Man Cybern. 41(6), 797–819 (2011)
7. Luxand Inc.: Luxand FaceSDK 4.0. http://www.luxand.com/facesdk/ 2014. Accessed 2 Apr 2014
8. Ekman, P., Friesen, W.: Unmasking the Face: A Guide to Recognizing Emotions from Facial Clues. Prentice Hall, Englewood Cliffs, NJ (1975)
9. Kanade, T., Cohn, J.F., Tian, Y.: Comprehensive database for facial expression analysis. In: Proceedings of the 4th IEEE International Conference on Automatic Face and Gesture Recognition, pp. 46–53 (2000)

A Procedure for the Development of Mobile Applications Software

Byeongdo Kang, Jongseok Lee, Jonathan Kissinger and Roger Y. Lee

Abstract In this paper, we present a development method for mobile applications. Mobile applications are different from desktop applications because mobile devices are resource-limited embedded systems and their application domain is smaller than desktop domain. So we need a development method that is simply applicable to mobile software running on mobile devices. Our method consists of five development phases including requirements analysis, architecture design, navigation design, page design, and implementation and testing phase. Software developers can apply our method to developing their applications in a straightforward manner. We applied this method to developing an information system for locating a classroom. Our result were quite promising and warrant further research.

Keywords Mobile applications software · Software development · Development method

1 Introduction

Applications running on mobile devices are becoming so popular that they are representing a revolution in the IT sector [1]. A mobile platform consists of various source codes to control a microprocessor and hardware. Mobile applications include various

B. Kang (✉)
Department of Computer and Information Technology, Daegu University,
Daegu, Republic of Korea
e-mail: bdkang@daegu.ac.kr

J. Lee
Department of Computer Engineering,
Woosuk University, Wanju-gun, Republic of Korea
e-mail: jong1007@ws.ac.kr

J. Kissinger · R.Y. Lee
Department of Computer Science, Central Michigan University, Mount Pleasant, USA
e-mail: kissilj@cmich.edu

R.Y. Lee
e-mail: lee1ry@cmich.edu

© Springer International Publishing Switzerland 2015   141
R. Lee (ed.), *Software Engineering Research, Management and Applications*,
Studies in Computational Intelligence 578, DOI 10.1007/978-3-319-11265-7_11

software that uses APIs supported by a mobile platform [2]. This paper presents a development method to help software engineers to improve productivity and quality of mobile applications.

In Sect. 2, we present general background information regarding mobile applications development approaches. In Sect. 3, we introduce the characteristics of mobile applications. Then we present a development method in Sect. 4. We apply our method to an application example in Sect. 5. Finally we come to a conclusion in Sect. 6.

2 Related Methodologies and Development Approaches

There are many preexisting approaches to developing mobile applications, in this section we are going to discuss some of the more notable ones and compare them to our approach.

Kraemer [3] details his approach which focuses on limiting complexity and development time. To that end he suggests the use of libraries and tools that are highly extensible. These tools successfully reduce development time and complexity, especially for novice programmers or those new to mobile development. However, third party tools carry inherent risks in addition to their benefits. They sacrifice reliability because they are not supported and can break easily. As Kramer notes, mobile platforms release new versions often. Even with careful consideration this can lead to instability, especially in tools that are extensible. Because of this we suggest a method that does not rely on third party tools or libraries.

Dehlen and Aagedal [4] propose a unique mobile deveolpment method that focuses on mobility. Their proposition is that mobile applications should be aware of the devices movement at the physical, logical, and network level. Many applications would benefit from such an architecture, however not all applications would. Self-contained application, those not requiring network access at all, as well as applications that require little location awareness such as solely GPS, would see little, if any, benefit from such an approach. Thus, this method will add unnecessary complexity to many mobile applications. Whereas our approach is heavily focused on simplicity, and universal applicability. Finally in most modern mobile platforms the operating system handles all location information. Applications need only query the operating system to acquire the necessary information. Rarely if ever do applications need to manipulate such information since it's handled by the OS.

Another useful approach given by Parada and Brisolara [5] provides specifications for developing Android applications. This approach use models to solve the problems of code efficiency and development time. Similar to the approach of Kraemer [3] this method uses code generation and tools. Which, while extremely useful, can lead to issues with reliability. Also this approach focuses solely on the Android operating system and is not an adequate approach for all mobile platforms.

A fourth approach presented by Balagtas-Ferrnandez and Hussman [6] explains how novice programmers may be put off by development because of the inherent level of difficulty. To solve this problem they have created a tool called the MobiA modeler [6]. A cross-platform tool which is excellent for beginning developers, the MobiA however has the same pitfalls as previously tools, specifically with Kraemer as well as Parada and Brisolara's approach. Because the tools are platform agnostic however, they prevent the user from having a native experience on their device, something which even web applications struggle with [7].

Given the rapid nature of mobile platform evolution and the lack of formal research into mobile engineering technologies, it should be no surprise that new and varied approaches to development are necessary [8]. Our approach is unique in that it focuses on simplicity without sacrificing reliability, and universal applicability without sacrificing a native user experience. We also expect our approach to have more longevity than other methods due to the high level nature.

3 Characteristics of Mobile Applications

There are two kind of traditional mobile applications: web-based applications and native applications. Web-based applications consist of web pages optimized for mobile devices and can be developed by using HTML, JavaScript and CSS. They usually run on a server. Native applications are developed for specific mobile devices. They can access the functionalities of mobile devices, such as GPS, file storage, databases, SMS, mail box and etc. They can be downloaded, installed, and sold in applications store [9]. In this paper, mobile applications refer to native mobile applications.

Mobile devices, in particular smart phones, have been popular in our life. Mobile applications are different from the desktop applications because smart phone devices are resource-limited embedded systems and are developed separately on different development platforms with different operating systems such as RIM of Blackberry, Windows Phone, iOS, Symbian, and Android [10]. In addition, mobile applications are developed in small-scale, fast-paced projects to meet competitive market demand [11].

Development of mobile applications with Android requires Java programming language using the Android SDK which provides the tools and APIs necessary for development [12]. An application is packaged as an Android application. Figure 1 shows the Android applications architecture.

4 A Development Method for Mobile Applications

We propose a development method for mobile applications, which includes five phases: requirements analysis, architecture design, navigation design, page design, and implementation and testing. Figure 2 shows the entire development cycle. This process model is iterative between phases to support feedback.

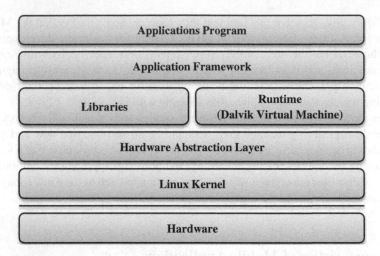

Fig. 1 Android system architecture

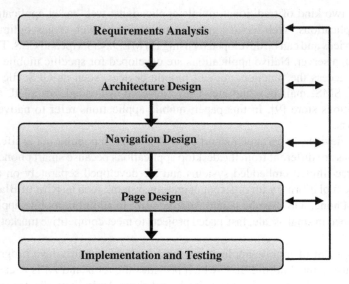

Fig. 2 A development phases for mobile applications

4.1 Development Phases

Requirements Analysis. Developers define the goals and functions of the mobile applications. The purpose of the requirements analysis phase is to analyze the application domain through the viewpoint of users. In this phase, developers define the target users who will use the application. They also analyze the contents and functions, constraints, and who is going to provide the new content. The product of this phase is requirements specification.

Architecture Design. Developers determine the most suitable architecture according to the result of the requirements analysis phase. Developers divide the application domains into sub-applications. Well-defined architecture can reduce the complexity of the system and provide the work boundaries for developers. The product of this phase is the architecture design diagram.

Navigation Design. Developers define navigation relationships between pages (screens of smart phones) of the mobile applications. The navigation relationship includes the link relationship and data migration between the pages and makes the mobile applications different from general applications. The mobile applications program generally consists of more than one page. Users of applications navigate the pages to retrieve some information or to accomplish what they want to do. The product of this phase is the navigation design diagram.

Page Design. Developers design the screen layouts and functions for all of the pages. The pages can be classified into static pages and dynamic pages according to their functions. The function of static pages is to show their contents. The function of dynamic pages is to accomplish tasks such as data processing or accessing databases. The products of this phase are the page detail design diagrams.

Implementation and Testing. The analysis and design specifications can be implemented in a straightforward manner by programming all of the page detail design diagrams. The behavior of the mobile applications must be tested on the emulator and on the mobile device because the applications on an emulator may perform differently from them running on a mobile device with various hardware and software versions [13].

4.2 Graphic Notations

Our method provides program developers with two main notations, the component and the connector. Components represent the functional modules of the system while connectors represent the interactions between components. Figure 3 represents the notations of the diagrams for modeling mobile applications

Components in the diagrams are classified into the architecture component, the page component, the passive component, the active component, the database component, the group component, and the condition component.

The architecture component is used to represent the structure of applications in the architecture diagram and represents a function of an application. The page component represents a page in the navigation design diagram. The passive component represents a static functional module. The active component represents a dynamic module. The database component represents a data repository. The group component can be used to combine a set of components into one group of functions. The condition component is used to specify a condition. All of these components are used in the page detail design diagram.

Components   Connectors

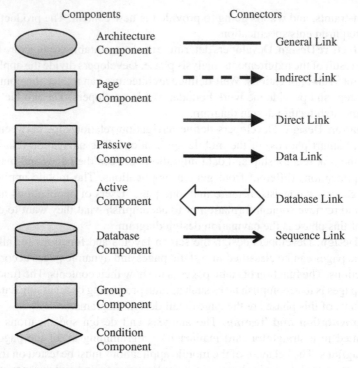

Architecture
Component   General Link

Page
Component   Indirect Link

Direct Link

Passive
Component   Data Link

Active
Component   Database Link

Database
Component   Sequence Link

Group
Component

Condition
Component

Fig. 3 Graphic notations for design diagrams

Connectors in the diagrams are classified into the general link, the indirect link, the direct link, the data link, the DB link, and the sequence link.

The general link represents the existence of any relationships between two components in the architecture design diagram. The indirect link and the data link represent the transitions occurred by a user's clicking on a button. The indirect link does not contain a data transmission between two components. But the data link contains a data transmission between two components. The direct link represents an automatic page link in the applications program. The Database link represents a data transmission between a functional module and a database. The sequence link represents the sequence of the activation of components. The Database link and the sequence link are used in the page detail design diagram.

4.3 Diagrams

The following three kinds of diagrams are produced after we analyze and design an application:

• The architecture design diagram

- The navigation design diagram
- The page detail design diagram

The Architecture Design Diagram. The architecture of software is defined by computational components and interactions among components. The well-defined structure makes it easy to integrate and maintain the parts of a large application. The architecture design diagram shows the vertical and horizontal structure between functions of applications and does not include the information about the detail algorithms.

The Navigation Design Diagram. The most important characteristic of mobile applications is the navigation feature. Because mobile applications consist of pages (for example, one screen form of a smart phone), users of the applications have to explore pages to search for information or accomplish what they want to do.

The navigation design diagram represents the navigation relationships among pages. It shows the link relationships and data transformation among pages.

The Page Detail Design Diagram. The page detail design diagram represents each page in detail. The pages are classified into static pages or dynamic pages according to their tasks. Some pages may include the characteristics of both.

The static pages display their contents and are described by the design patterns. On the other hand, the dynamic pages perform some tasks and are described by the functional flows to represent the algorithms for the tasks.

5 Mobile Application Example: Location a classroom

We apply our method to developing a mobile application that provides the information about the locations of classrooms. It requires a classroom number as an input from users, and then shows the information about the location of the classroom. Figure 4 shows the application development environment supported by Android and Eclipse.

Figure 5 shows the architecture design diagram for the main starting page. Here you can see the simplicity and high level nature of our method. The application includes two computational components: Search and Exit. The component Search is processed by the two components, Success and Failure, checking the validation of the input from users. The architecture design diagram includes architecture components and general links.

If users enter a correct classroom number, the information system prints its location. If they don't, it prints failure message. The component Exit includes the procedures for users' exiting the information system.

Figure 6 shows the navigation design diagram for the architecture design diagram in Fig. 5. Figure 6 represents the navigation relationships between pages for the application by using the page component, the direct link, the data link, and the database link. The component in this diagram is concretized in the page detail design diagram.

Figure 7 shows the page detail design diagram for Fig. 6. Figure 7a is the screen layout of the start page of the application. It shows users two button types, Search and

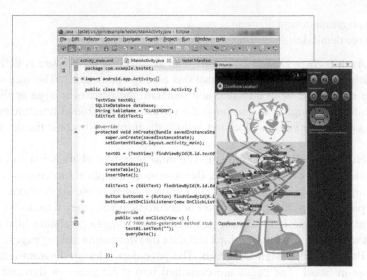

Fig. 4 Application program development environment

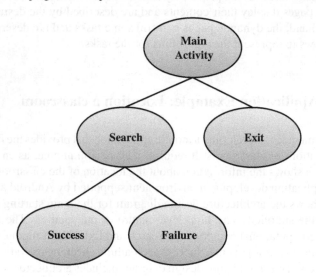

Fig. 5 The architecture design diagram

Exit, at bottom. These can be implemented differently for different OS allowing for a native user experience while maintaining a simple design diagram. The position of input that users enter a classroom number is above two buttons. The result of users' query is printed at top. A graphic image is displayed between the result and the input. Figure 7b represents the functional flows of the function "Search". It is operated by pressing the button Search.

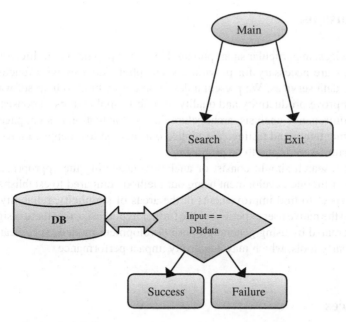

Fig. 6 The navigation design diagram

(a)   **(b)**

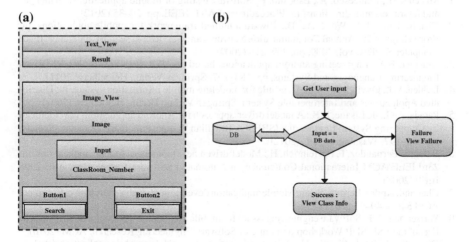

Fig. 7 The page detail design diagram. **a** The design patterns **b** The functional flows

Comparison of our approach with other mobile development models shows advantages in the areas of simplicity and elegance as outlined by the above figures and explanations. Also see Sect. 2 for more detailed information about the differences between our proposed method and preexisting methods.

6 Conclusions

Mobile devices, in particular smart phones, have been popular in our life. So, mobile applications are necessary for providing smart phone devices with functionalities for mobile data services. We present a development method to help software engineers to improve productivity and quality of mobile applications. It consists of five phases: requirements analysis, architecture design, navigation design, page design, and implementation and testing. We applied our method to an application example, and the results were found to be promising.

Further research should consist or analytics measuring the appropriate quality attributes for mobile development using our method compared to established methods. We expect to find improvements in the areas of simplicity, reliability, performance, and the native user experience . We further hypothesize that there is significant overhead created by using general purpose development models, specifically UML and third party tools, which may negatively impact performance.

References

1. Muccini, H., Francesco, A., Esposito, P.: Software testing of mobile applications: challenges and future research directions. In: Proceedings of AST, IEEE, pp. 29–35 (2012)
2. Cha, S., Kurz, J.B., Weichang, D.: Toward a unified framework for mobile applications. In: Proceedings of 7th Annual Communication Networks and Services Research Conference. IEEE Computer Society, vol. 2009, pp. 209–216 (2009)
3. Kraemer, F.A.: Engineering android applications based on UML activites. In: Model Driven Engineering Languages and Systems, pp. 183–197. Springer-Verlag, Heidelberg (2011)
4. Dehlen, V.J., Øyvind, A.: A UML profile for modeling mobile information system. In: Distributed Applications and Interoperable System. Springer-Verlag, Berlin, Heidelberg (2007)
5. Parada, A.G., de Lisane, B.B.: A model driven approach for Android applications development. In: Computing System Engineering (SBESC), Brazilian Symposium. IEEE Computer Society, pp. 192–197. Washington, DC (2012)
6. Balagtas-Fernandez, F.T., Heinrich, H.: Model-driven development of mobile applications. In: 23rd IEEE/ACM International Conference on Automated Software Engineering, ASE 2008 IEEE (2008)
7. Charland, Andre, Leroux, Brian: Mobile application development: web versus native. Commun. ACM **54**(5), 49–53 (2011)
8. Wasserman, A.I., Software engineering issues for mobile application development. In: Proceedings of the FSE/SDP Workshop on Future of Software Engineering Research. ACM (2010)
9. Divya, S., Nikita, J., Shruthi, U., Rajat, G.: Generic framework for mobile application development. In: Proceedings of the 2nd Asian Himalayas International Conference on Internet, IEEE, pp. 1–5 (2011)
10. Wei, H., Hong, G.: Curriculum architecture construction of mobile application development. In: Proceedings of International Symposium on Information Technology in Medicine and Education, IEEE, pp. 43–47 (2012)
11. Mona, E., Ali, M.: Reverse engineering iOS mobile applications. In: Proceedings of 19th Working Conference on Reverse Engineering, IEEE Computer Society, pp. 177–186 (2012)
12. Butler, M.: Android: changing the mobile landscape. IEEE Pervasive Comput. **10**:4–7 (2010)
13. Dantas, V., Marinho, F., Costa, A., Andrade M.: Testing requirements for mobile applications. In: Proceedings of ISCIS, IEEE, pp. 555–560, 14–16 Sept 2009

The Relationship Between Conversation Skill and Feeling of Load on Youth in Communicate with Elderly Persons Using Video Image and Photographs

Miyuki Iwamoto, Noriaki Kuwahara and Kazunari Morimoto

Abstract Conversation is a good preventative against behavioral problems in the elderly. However, caregivers are usually very busy tending to patients and lack the time to communicate extensively with them. In order to overcome such problems, active listening volunteers have more opportunities to communicate with the elderly, but the number of the skillful volunteers is limited. Therefore, conversational support systems for inexperienced volunteers was investigated; such systems usually include content such as photographs, videos and music. It was expected that the volunteers would feel less stress when using videos instead of photographs for conversational support because the former provided both volunteers and patients with richer information than the latter. On the other hand, the photographs gave patients more chances to talk with volunteers. However, there are no reports on the effect of content type on the stress and conversational quality. In this paper, we compared using photographs with those when using video form such viewpoints.

Keywords Elderly · Reminiscence videos · Senior care home

1 Background and Purpose

Japanese society has been recently facing an unprecedented problem-super-aging. Population aging has accelerated rapidly in Japan, and now the aging rate is very high; 35 million people are expected to be 65 years or older in 2018 [1].

At the same time, the proportion of households consisting of elderly couples or senior citizens living alone has also increased, along with the number of elderly

M. Iwamoto (✉) · N. Kuwahara · K. Morimoto
Graduate School of Engineering and Science, Kyoto Institute of Technology,
Kyoto, Japan
e-mail: cabotine.six.stars@gmail.com

N. Kuwahara
e-mail: nkuwahara@kit.ac.jp

K. Morimoto
e-mail: morix@kit.ac.jp

© Springer International Publishing Switzerland 2015   151
R. Lee (ed.), *Software Engineering Research, Management and Applications*,
Studies in Computational Intelligence 578, DOI 10.1007/978-3-319-11265-7_12

people spending their days without speaking, thereby leading to disuse of cognitive functions, risk of dementia, and depression.

However, in helping to communicate with the elderly, understanding their physical and psychological characteristics is required.

For example, it is necessary to deal with declines of cognitive functions, hearing, and vision, which damages elderly patients' self-esteem. Listening therapists, volunteers and clinical psychologists can interact with the elderly with the understanding that you have these. However, at present, personnel's lack overwhelmingly compared to the number of the elderly.

So, in the future, it is expected that the younger adults will be involved with the elderly as actively dialogue partners. However, it is a problem that younger adults are unfamiliar with the communicating with the elderly. In order to address this problem, conversation support systems using video and photographic images have been tried already. However, in any studies, it has not been made research about a feeling of burden that the younger adults as students or volunteers feels.

Until now, studies of interaction support for the QOL improvement of the elderly was main stream; the burdens of younger adults who communicate with the elderly were overlooked. However there has been no report on the effect of the type of conversation support content, i.e., if there is any difference of feeling of burden between image and movie, and whether there is a change in the quantity and quality of dialogue.

In our study, we took advantage of conversational support content in our attempts for the elderly, students, and volunteers to communicate. It was confirmed that smooth conversation is of value to the elderly but that even in these cases, many volunteers and students complained about feeling burdened while communicating (Fig. 1).

Many volunteers and students who are inexperienced in listening feel burden when conversing with the elderly, or that they cannot converse well with senior citizens in many cases. We focused on the conversation support in reminiscence for volunteers and students to reduce the burden on the conversation with the elderly. In addition, research into promoting conversations between young adults (volunteers and students) and the elderly was done by providing allowed topics via photos and video images. (Example [2–4]) Information medium for providing topic here (hereinafter referred to as "media") is referred to as conversation support content. On the other hand, the photographs gave the patients more chances to talk with volunteers. However, there have been no reports on the effect of the content types on the stress levels and conversational quality.

Therefore, in this study we measured the levels of emotional burdens which younger adults (volunteers and caregivers) felt when talking with elderly people while using videos and photographs as conversational support content. The purpose of this study was to verify the method of utilizing the best media as conversation assistance.

The experiments were carried out in research presented at the Human Interface symposium [6]. Young adults conversed with the elderly using both video and still images, and which conversational support content caused less burdensome feelings was noted; in order to evaluate conversational quality and quantity these experiments

Fig. 1 Volunteers utilizing conversation support content

were carried out with the 5-stage subjective assessment used in previous conversational support studies [2].

Further, the speech time by video was also evaluated in line-of-sight utterance. Because elderly person-student pairs only conversed once, it was found that measuring burdened emotions was difficult because of their individual differences, such as the characteristics of the elderly. Therefore, we measured the burden feelings by averaging the effects of such affinity levels with elderly in that it has carried out a conversation with the elderly of five for one student.

For conversation time, for searching the Internet and video image during a conversation, a problem folding the waist of the story, and that not be measured properly conversation time search takes too much time and even occurs. Therefore, eliminating the burden of the search operation by previously prepared image and video during a conversation, we measured the load of only the interaction of conversation.

For measurement items, we were pointed out that in the video shoot or questionnaire, subjective burden evaluation only may be made, in addition to the burden subjective evaluation, we measured the heart rate as an objective index. When students search for photos and video images on the fly from internet.

Content to be the choice of each student is different. And then when the content of interest and favorite things has been depicted, to compare the burden of conversation is difficult. Therefore, with the use of the same content depicted in the photographic image and video, and control the effects of the content in the quality rating of the conversation.

To improve these points, an experiment was conducted.

In Sect. 2, solved the problems of the previous experiment, enhanced the reliability of the data collected, an experiment was conducted on the conversation that was subject students and the elderly (including persons with dementia).

Describe the evaluation items and the experimental outline.

In Sect. 3, describes of the evaluation items consider and describe in detail on the experimental results.

In Sect. 4, describe the discussion of this study.

In Sect. 5, describe the challenges of the future.

2 Experiment

2.1 Summary

We conducted a questionnaire with the same protocol as the last time, but there was no significant change when compared to previous results. We showed the various degrees of burdensome feelings, heart rate, number of utterances and in the line-of-sight this time.

In previous experiments we also included entertainment topics as a category, but this was considered inappropriate as the difference in conversational content the age for young adults to the elderly is large. We determined, that "food", "event" and "play" would be the conversational support content, based on the basis of previous studies.

From how to improve we have measured the burden by averaging the effect of such affinity with the elderly in that it has carried out a conversation with the elderly of five for one student.

For speech time, by eliminating the burden of the search operation by previously prepared image and video during a conversation we measured the load of only the interaction of conversation

For measurement items in addition to the subjective burden evaluation levels, we also measured the heart rate as an objective index. By using the same representational content for photos and videos, we controlled the contents' effects in the conversational quality rating.

For evaluating conversational quality, we carried out in each conversation the questionnaire depending on the fifth stage subjective evaluation.

For the evaluation of conversational amounts, we evaluated speech time and the line-of-sight of the utterance from video obtained by photographing the conversation.

2.2 Evaluation Items

We carried out checks on the degree of burden of conversation by students. A stress check board was positioned out of sight of the elderly patients (Fig. 2). Numbers "1–7" was written on this stress check board, "1" equating to no (stress) burden during the continuation of the conversation, "7" meaning that lots of stress was felt. At the beginning of the experiment, students' stress levels were at "1". The students' stress levels were measured every minute. At that time, students chose "1–7" according to the degree of stress they experienced during a conversation.

It represents the degree of stress in conversation at that time. During the conversation, the expressions (appearance and nods) of students and the elderly were

Fig. 2 Stress check borad

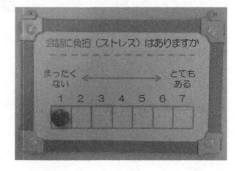

recorded at that time. We also measured the heart rate in order to determine the students' conversational stress levels.

During the evaluation of the conversational quality, a questionnaire was given once during each experiment according to the fifth stage subjective evaluation.

2.3 Experiment Environment

Figures 3 and 4 show the layout and appearance of the experimental environment.

The experiments were carried out in one room at the facility's nursing home. The subjects made conversation while sitting in chairs placed side by side. We used a desktop PC in which video and photographic images were previously stored by category. The Stress Check Board was placed where the elderly subjects couldn't see it, and the students pointed at a number in the direction of the experimenter.

2.4 Subject Information

The subject students were two (23-year-old) college students. They were diagnosed in advance with the Yatabe-Guilford sociability personality diagnostic test.

Fig. 3 Experiment environment

Fig. 4 The layout of the experimental environment

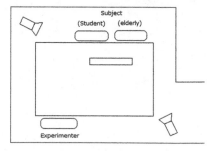

Eight elderly women in their 80s–90s were residents in nursing homes suffering from mild dementia.

2.5 Experimental Procedure

It was assumed that elderly and students would converse while looking at previously-prepared photo images and videos. About 20 still images depicting each topic—"food","event","play"—were prepared in advance. It was assumed that the students would choose the next photo that fit the conversation. The conversation lasted 10 min.

The students pointed to the stress check board the burden degree of continuity of the conversation on a minute-by-minute. Stress check board to be installed in a position that is not visible to the elderly. In this study, rather than evaluating the complaints about the conversationalist's partner, checking the conversation's burden levels conversation was the object; without relying on the conversation partner, pure feelings were recorded.

Questionnaire evaluations were conducted during each experiment according to the fifth stage subjective evaluation. In order to measure the students' conversational stress, their heart rates were also measured.

3 Result

The experimental results are shown below. Figure 5 shows the stress accumulated amount of both students A and B. The horizontal axis represents the photos and videos of subjects A and B, respectively, and the vertical axis represents the accumulation on the numerical stress checkerboard in Fig. 5. This is a comparison of the younger adults' stress levels in the case of photo usage, in the case of video usage as a conversation support content feels continuation of the conversation.

Fig. 5 Integration of the 10 min of degree of burden

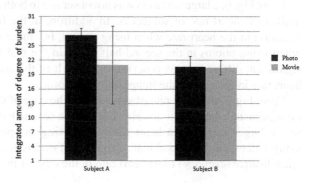

Fig. 6 The first half of the subject A, the average of the heart rate of the second half

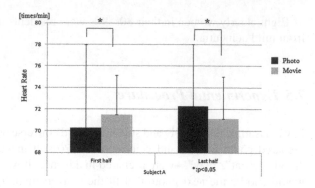

Fig. 7 The first half of the subject B, the average of the heart rate of the second half

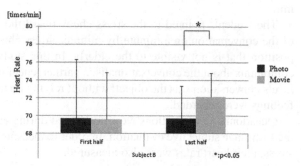

The two subjects, A and B, are respectively shown below. From the results of the personality test, Subject A scored 58 % sociability, 50 % cooperativeness. On the other hand, Subject B scored 75 % sociability, 61 % cooperativeness.

Subject A and B showed an observed difference in sociability. In subject A, using photo images tended to result in higher stress levels.

Further, we measured the first half of the heart rate and the average of the second half during conversations. In Figs. 6 and 7, the vertical axis indicates the heart rate [times/min] of students during conversation. The horizontal axis represents the first half of the experiment time, the second half.

From Fig. 6, a large variation was not observed in both the first half and the second half of the heart rate of subject A. In addition, in the first half, there was a large variation in the heart rate when using videos, but there was also a large difference when using photos in the second half. Both in the first half and the second half, a significant difference was observed as a result of the (significance level 5 %) t-test heart rate in the case of the image and the photo.

From Fig. 7, the heart rate of subject B in the case of the video and photo showed no significant change initially. However, in the second half of the conversation, there was a larger variation in heart rate when using the photos than there was when using videos. In the second half, a significant difference was observed result of the (significance level 5 %) t-test heart rate in the case of the video and photo.

Fig. 8 Speech time
(subject A)

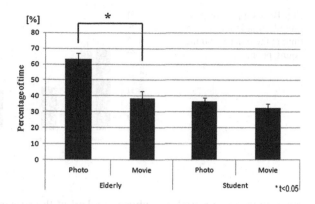

Fig. 9 Speech time
(subject B)

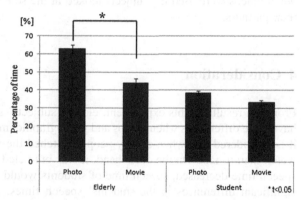

Figures 8 and 9 shows the results obtained by aggregating the conversation times of young adults and the elderly from video that was recorded during the experiment. The horizontal axis shows the photo and video usage of elderly and students each. The vertical axis shows the proportion of speech time of 10 min in the experimental time (%).

From the results of Figs. 8 and 9, it can be said that the amount of speech has become a lot at the time of the video than photo for the elderly and younger adults both. In the case of the elderly, significant difference were observed in speech times for both the video and photo. On the other hand, in the case of young adults, significant difference were not observed in the speech times for photo. For both the elderly and young adults, a significant difference was observed result of the (significance level 5 %) t-test the speech time in the case of both video and photo.

Further, we were compared of the line-of-sight of the utterance time (Fig. 10). Line-of-sight in this study were shown in the graph; when subjects were looking at the screen and when they were looking at their conversation partners are also indicated. When they were looking at the photo, it tend that both the elderly and students looked at each other more than they looked at the screen. On the other hand,

Fig. 10 Comparison of the line-of-sight in the speech time in the elderly and the young people

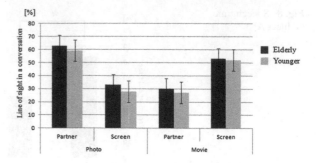

when videos were used the subjects looked at the screen more than they looked at their partners.

4 Consideration

From the results of this experiment, elderly subjects' speech times were found to have large differences when videos and photographic images were used. As a cause, it is considered that when elderly people watched the video, it would be larger to watch than to talk. On the other hand, it was predicted that if the elderly subjects' speech time decreased, speech time of students would increase, but there were no significant differences in the students' speech times. In addition, when both the elderly and students used photos, it was found that their speech times increased. The reason why the speech times of the elderly were longer in the case of photographic images was discovered in the video analysis results.

In the case of photographic images, when the elderly gazed at the image, their speech times were very short. After that, the conversation was not related to the photograph that followed. On the other hand, in the case of the video, the time spent watching the video was long, and the elderly spoke unilaterally less. As a result, it is believed that for students, conversation became more lively, and with a picture it was hard to experience burden feelings during the conversation.

In this study, in the results of speech times, when using photos with both elderly and students discovered that speech times were long. Also from the length of the speech time, even from the subjective evaluation of the previous paper [4], it was considered that the by using photos, individuals can have valuable conversations with the elderly. On the other hand, from the previous paper [6], using video resulted in lowered burdened feelings in the subjective evaluation of students. In other words, I considered the length of the speech time and was able to talk intimately.

5 Challenges for the Future

Future, it is necessary to verify when young adults feel burden and to search burden of speech, the speech content and whether feel of the burden. Therefore, in accordance with young adults' burdened feelings, needing to construct a system which can be used by switching the appropriate conversation support from video to photographic images will be a problem.

At this time, as noted in a previous paper [4], while continuing the conversation is difficult to come up with the search words a specific difficulty in operability in order to solve. In particular, using search words, finding the locations of a myriad of images, searching YouTube, and by providing a glossary of more detailed topics such as "food", "events", and "play" reduced the students' burdensome feelings. Further, it is consider that to build such a system can search the image or video by utilizing the touch panel.

Acknowledgments This work was supported by JSPS KAKENHI 24650037 A part of this research was supported by COI Stream of MEXT

References

1. A Overview Ministry of Health, Labour and Welfare, a 12-year Heisei version of Annual Report on Health and Welfare, the aging of the world, the Ministry of Health, Labour and Welfare website (online). http://www1.mhlw.go.jp/wp/wp00_4/chapt-a5.html, detail_recog.html (Reference 2012-01-13)
2. Astell, A.J., Ellis, M.P., Bernardi, L., Alm, N., Dye, R., Gowans, G., Campbell, J.: Using a touch screen computer to support relationships between people with dementia and caregivers. Interact. Comput. **22**, 267–275 (2010)
3. Kuwahara, N., Kuwahara, K., Abe, S., Susami, K., Yasuda, K.: Video memories that utilize annotation of photo-Application and evaluation to persons with dementia—making support. Artif. Intell. J. **20**(6), 396–405 (2005)
4. Tuji, A., Kuwahara, N., Morimoto, K.: Implementation of interactive reminiscence photo sharing system for elderly people by using web services. Human Interface Society, Kyoto (2010)
5. Takai, S.: Topic suggestions propulsion system of conversation during the first meeting in accordance with the TPO, Heisei 21 year master's thesis (2010)
6. Iwamoto, M., Kuwahara, N.,: Morimoto, K.: Comparison between the Burden of the conversation by using photographic image and that by using motion video. Hum. Interface Soc. **2012**, 579–584 (2012)

Automatic Matching on Fracture Surface of Quarried Stone Using the Features of Plug-and-Feather Holes

Karin Morita, Sei Ikeda and Kosuke Sato

Abstract In archeology, automatic matching of quarried stones is important for investigating their processing steps and distribution channels, and can save large amounts of human efforts. In this paper, we propose a novel method for matching of quarried blocks of a stone based on feature extraction of plug-and-feather holes, which are holes drilled for splitting the original stone. The cues for detecting holes are the unique shape of plug-and-feather holes. The positional relationship of holes is used to estimate the initial position for the alignment of fracture surfaces. Automatic matching is achieved through the alignment of the fracture surface by ICP algorithm from its initial position. The experiments for an actual stone show the advantage of feature detection in automatic matching of quarried stones.

Keywords Plug-and-feather hole · ICP algorithm · Automatic matching

1 Introduction

Matching of quarried stones which have the plug-and-feather holes is one of important tasks for investigation of the split order and distribution channels in archeology. Plug-and-feather holes, also called wedge and shim holes, are formed in a stone splitting process. In this process, holes are drilled on a stone surface in line, and plug and feathers are inserted into each hole. The fracture surfaces of such stones have few unevennesses and geometric features compared to those of naturally cracked relics, since the stone is divided in the direction in which the stone easily splits. Traditionally, matching operation of quarried stones has been performed manually. In manual matching, correspondences should be found from geometric features and

K. Morita (✉) · S. Ikeda · K. Sato
Osaka University, Toyonaka, Osaka, Japan
e-mail: morita@sens.sys.es.osaka-u.ac.jp

S. Ikeda
e-mail: ikeda@sys.es.osaka-u.ac.jp

K. Sato
e-mail: sato@sys.es.osaka-u.ac.jp

© Springer International Publishing Switzerland 2015   163
R. Lee (ed.), *Software Engineering Research, Management and Applications*,
Studies in Computational Intelligence 578, DOI 10.1007/978-3-319-11265-7_13

texture on fracture surfaces. However, a great deal of time and effort is required for manual matching process.

Automatic matching for multiple range images acquired from different viewpoints have been proposed. Existing alignment methods for three-dimensional (3-D) surfaces can be divided into two types: Iterative Closest Point(ICP) based methods and feature point-based methods. In ICP algorithms, the alignment is performed by repeating the follow two processes. First, corresponding points with minimum distances are searched. Next, the rigid transformation which minimizes the sum of the distances are estimated. However, an alignment by an ICP algorithm strongly depends on the initial position and a shape of the object because there can be many local optimal solutions around the global one especially in the case of semi-planar surfaces [17] such as fracture surfaces of quarried stones. It is necessary to find a good initial position before performing the ICP algorithm.

Feature point-based methods detect feature points, and directly estimate the optimal rigid transformation which minimizes the sum of distances between corresponding pairs. NARF [18] is a technique to extract the feature based on the information of normal from the 3-D range data. Spin-image [9] is a descriptor image of the surface shape around the feature points and the correspondence between the feature points are computed on the basis of the covariance. The result of alignment by a feature-point method does not depend on the initial position. Since each plug-and-feather hole consists of multiple faces, feature points tend to be detected near the edges. However, the feature values between corresponding points may not match since the feature values computed at the feature points can include other part information than fracture surfaces. On the other hand, the feature value between feature points on the parts of the surface have small differences since fracture surface are flat. In addition, some fracture surfaces do not match since they can be partly carved and flattened.

In this paper, we propose a novel matching method using the feature of plug-and-feather holes. We utilize the unique shape and position of the plug-and-feather holes as cues of corresponding faces. Plug-and-feather holes have flat parts whose sizes are almost constant.

The proposed method consists of three processes: detecting the plug-and-feather holes, estimating initial position and aligning fracture surfaces by a multi-resolution ICP algorithm. In order to confirm the effectiveness of the proposed method, we first show that matching of quarried stones only using ICP algorithm is difficult, then we show a result of automatic matching by using the proposed method.

2 Related Work

2.1 Acquisition of Data

The proposed method assumes that the whole 3-D shape data of divided stones are given. The most typical method in a non-contact manner is using laser scanners for a large-scale environment [7]. However, stationary devices are not suitable for complex target, since frequent movements of devices are needed in order to prevent occlusions.

On the other hand, the above problem does not occur in methods using hand-held sensors such as a digital cameras and depth sensors. 3-D reconstruction from multiple images computes the 3-D positions of each point by the principle of triangulation through determining corresponding points between images obtained from various viewpoints. SLAM typified by DTAM [11] can estimate a dense surface in real time. However, computational cost is still high. Methods using the depth sensor can perform 3-D reconstruction in real time by aligning the depth image frames while moving the sensor relative to the object [15]. This technique has characteristics that measurement is possible without depending on texture of the target object and reconstruction result can be visualized on site. In this research, we performed measurement using "SENS-Wiper" [8]. SENS-Wiper is one of the above methods using depth sensors. SENS-Wiper is a wearable shape measurement system that can measure outdoors, and consists of Kinect, laptop PC and battery.

2.2 Alignment of Range Images

Registration of incomplete 3-D shape data obtained from various viewpoints is required in order to obtain a complete shape data of the object. The relationship of viewpoints is easy to acquire by using a turntable for a change of viewpoint [1]. However, in outdoor measurements, it is not easy to obtain precise positions of sensors in most cases.

ICP algorithm [2] and feature point-based methods are often used for registration. ICP algorithm is a fine registration approach for 3-D shapes. The original method has problems that alignment cannot be performed correctly if the shape has a small characteristic part compared with the whole size of the object or if the shape has any portions without correspondences. Derivation techniques have been proposed to overcome this drawback. Chen and Medioni [4] proposed to use a distance from each source point to the plane (point-to-plane), has a better convergence characteristic than the point-to-point [17]. The method of Zhang [20] removes the corresponding points that there is little possibility of matching by the distance between corresponding points. Masuda's study [14] proposes a method to create an integrated shape from multiple range images to reduce accumulative errors. However, it is difficult to estimate the global optimal alignment if the viewpoint position is for from each other in ICP algorithm.

The feature point-based methods [9, 18] compute the optimal solution by using RANSAC based robust estimation in finding the similarity between descriptors of the feature points. Spin-image [9] is a descriptor generated by projecting the relative position of 3-D points that lie on the surface of an object to a 2-D space. NARF [18] detects the position of the corner from a depth image and describes direction of the corner as a feature.

2.3 Alignment of Fracture Surfaces

The alignment methods mentioned above are designed for multi-view depth images of the same object surface. In such multiple depth data sets, there must be corresponding parts except differences due to occlusions or noise. Matching fracture surfaces of quarried stone has further difficulties. The two fracture surfaces do not always match each other owing to cavities and artificial grinding of the convex parts. For these problems, there are methods based on continuity of the texture [12], symmetry of the original object [19] and geometric features excluding fracture surface [6]. However, it is difficult to apply these features to quarried stones since quarried stones do not have the above features.

3 Automatic Matching Method Using Plug-and-feather

3.1 Proposed Method Outline

The proposed method performs automatic matching from arbitrary initial positions to target two stones that have a sufficient number the plug-and-feather holes and has no texture pattern across the fracture surface. The 3-D shape data of the quarried stones as input data can be represented as a pair of point cloud: one called a data shape $D = \{D_i\}$ $(i = 1, \ldots, n)$ and the other called a model shape $M = \{M_j\}$ $(j = 1, \ldots, m)$. We compute the transformation T to state the data shape is aligned from the input position. T is the transformation matrix consists of a 3×3 rotation matrix R and a translation vector t:

$$T = \left(\begin{array}{c|c} R & t \\ \hline 0\ 0\ 0 & 1 \end{array} \right) \tag{1}$$

Figure 1 shows the data flow of the proposed method. The proposed method consists of three processes: First, the plug-and-feather holes are detected by a region growing method using normal vectors. The cues of the detection are the planarity and the size-constancy of the holes. Second, the initial position is estimated using the positions of the detected holes. Finally, we align the fracture surfaces using a multi-resolution ICP algorithm.

3.2 Detecting Plug-and-feather Holes

In order to estimate the initial position, the proposed method detects the plug-and-feather holes as feature points from the point cloud. For preprocessing, the raw point cloud data, the data is smoothed by the moving least squares (MLS) method [13].

Fig. 1 Data flow of the
proposed method

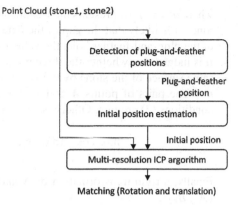

Matching (Rotation and translation)

After the smoothing, the plug-and-feather hole positions are detected by a region growing method based on the angle of the normal [16]. In this process, positions of the plug-and-feather holes are detected by using normals on the basis of the feature that the area of the hole is constant and flat as compared with other portions of the stone surface. The normals are calculated from the smoothed point cloud. We judge equivalence of each region by angle between normals and perform labeling using k-d tree.

3.3 Initial Position Estimation Using the Plug-and-feather

The initial position in the alignment is computed using the positions of plug-and-feather holes detected in the previous section. A pair of corresponding plug-and-feather holes have the same positional relation Firstly, three point sets are selected randomly on each surface. After that, the transformation of the data shape is calculated so that the sum of the Euclidean distances between the corresponding points is minimized. Since correspondences between the positions of holes are unknown, candidates of correspondences are selected using the direction of the normal and the distance between points.

1. As candidates, pick out three points of the plug-and-feather holes on each surface randomly. Points selected from the data shape are defined as $A = \{A_i\}$, and points selected from the model shape are defined as $B = \{B_j\}(i, j = 1, 2, 3)$.
2. Remove the combination of point cloud A and B, and return to step 1 unless it meets the following condition (2).

$$\max_{i \neq j} |d^u(A_i, A_j) - d^u(B_i, B_j)| < \varepsilon \ (i, j = 1, 2, 3, i \neq j), \qquad (2)$$

where $d^u(X, Y)$ represents the Euclidean distance between two points, X and Y. Since it is likely that the gap of the detected positions is in the range of the size of the plug-and-feather hole, the value of threshold ε is the size of the hole.

3. It is judged that whether the three points of the point cloud A are present on the same surface of the stone by the value of the inner product of the normal vector n_X. The pairs of points A and B is kept as a the candidates if the following conditions (3) are met. Otherwise, return to step 1.

$$\max_{i \neq j} \cos^{-1}(n_{A_i} \cdot n_{A_j}) < \tau \ (i, j = 1, 2, 3, i \neq j) \tag{3}$$

4. Finally, a rigid transformation of A and B is computed by minimizing $\sum_k d^u$ $(A_k, B_k)^2$.

3.4 Multi-resolution ICP

Given the input point cloud with the estimated initial position, the transformation T that aligns the surfaces from the initial position is calculated. Calculation cost of ICP algorithm increases according to the number of point. Therefore, to reduce its computational cost, we apply multi-resolution ICP algorithm [10]. Multi-resolution ICP algorithm is a method which performs registration with down sampled data at first, and increases the resolution of the data in the subsequent iteration. The method reduces computational cost to judge the reliability of the initial position efficiently.

Figure 2 shows the procedure of the ICP algorithm for each stage of the multi-resolution ICP algorithm. Since the points belonging to inside surface of the

Fig. 2 Flow chart of the ICP algorithm

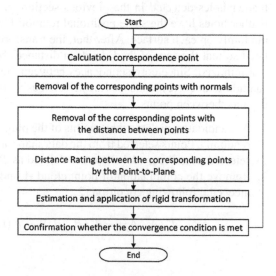

Fig. 3 Alignment results
depend on the corresponding
points used. **a** When all
corresponding points are
used. **b** When all
corresponding points (*green*)
are used

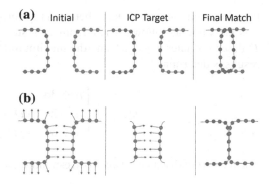

plug-and-feather hole or other than fracture surface are the erroneous correspondence, it is necessary to remove these points from corresponding points by using conditions described below.

3.4.1 Removal of the Unnecessary Points by the Normal Condition

We use corresponding points that the erroneous correspondence are removed since the ICP algorithm is affected by all points of the data shape and associates with the nearest point of the model shape. For example, when the corresponding points on other parts than fracture surfaces are used for alignment, the result should be incorrect as shown Fig. 3. When coarse alignment is achieved after the initial position estimation, we remove the unnecessary points similar to the method of Brown et al. [3]. We can remove the set of erroneous corresponding points, using the assumption that the normals of the pair of corresponding points face each other as shown (b). Specifically, the corresponding points are removed when the points do not meet the following condition using a threshold N of angle. We define $\boldsymbol{n}_{D_i}, \boldsymbol{n}_{M_i}$ as the normal vector of corresponding point D_i, M_i.

$$\cos^{-1}(\boldsymbol{n}_{D_i} \cdot \boldsymbol{n}_{M_i}) > N \tag{4}$$

3.4.2 Removal of the Unnecessary Points by the Distance Condition

On the surface of the stone, there is area does not contact each other even if normals face each other. For example, a part of the area that the feather contacted when stone was split. In this process, the correct corresponding points survive by removing the false correspondence. We remove such points with distances between the corresponding points as the study of Zhang et al. [20]. When distance between the corresponding points is greater than the maximum allowable distance D, the set of corresponding points are not used for rigid transformation estimation. D_{max}^I denote the maximum tolerable distance between the corresponding points in iteration I, and sets the value as follows. The alignment result of the iteration I with a threshold D described later is judged, and the maximum tolerable distance depending on the results is determined.

$$D_{max}^{I} = \begin{cases} \mu + 3\sigma & (\mu < D) \\ \mu + 2\sigma & (D \le \mu < 3D) \\ \mu + \sigma & (3D \le \mu < 6D) \end{cases} \tag{5}$$

The mean μ of distance between corresponding points and deviation σ are denoted by

$$\mu = \frac{1}{N} \sum_{i=0}^{n} d_i^u(D_i, M_j) \tag{6}$$

$$\sigma = \sqrt{\frac{1}{N} \sum_{i=0}^{n} (d_i^u(D_i, M_j) - \mu)^2} \tag{7}$$

where D_i and M_i denote a pair of corresponding points. As with the study of Zhang et al., D is the density of the data shape. Assuming that target point k_i of the data shape and n_k-point near the point are k_j' $(j = 1, 2, \ldots, n_k)$, the density ρ_i around the target point k is represented by the following equation.

$$\rho_i = \frac{1}{n_k} \sum_{j=0}^{n_k} d_j(k, k_j') \tag{8}$$

We choose 30 target points k_i randomly, and D is average value of the density ρ_i. After removing the unreliable relationship, we estimate the transformation using a set of remained corresponding points.

3.5 End Condition

Until the termination condition as shown below is satisfied, we repeat initial position estimation using the plug-and-feather holes and alignment with the multi-resolution ICP algorithm. The P_{in} is the probability that contains only inlier in the chosen sample, and γ is the number of current trials. The trial is stopped when the following equation is satisfied like Chum's method [5].

$$(1 - P_{in})^{\gamma} < threshold \tag{9}$$

P_{in} is determined as follows. n_d is the number of points of data shape, n_m is the number of points of model shape, and n_c is the number of corresponding points at the time of finishing the multi-resolution ICP algorithm.

$$P_{in} = \frac{2n_c}{n_m + n_d} \qquad (10)$$

4 Experiment

To demonstrate the effectiveness of the proposed method, we measured quarried stones with SENS-Wiper [8], which is a wearable 3-D shape measurement system. Figure 4 shows 3-D shape data of stone 1 and stone 2 measured by SENS-Wiper; the long-side of rectangle fracture surface is around 1,500 mm. Some archaeologists suspect that these two stones are from one stone.

Figure 5 shows the result of automatic matching in below setting. Threshold of angle in the region growing method was determined experimentally as 3.5°. The resolution of multi-resolution ICP algorithm was set to three layers. We used the data sampled to 1/100 in the 1st stage, 1/40 in 2nd stage and 1/20 in 3rd stage. Threshold of the angle of normal used in the corresponding point removal was 150°, the threshold of end condition was 0.01. The same values were applied in experiments described below. Figure 5 shows that irregularities on the surface of the stone 1 and stone 2 are continuous roughly. It was confirmed that the proposed method is able to automatic matching quarried stones from this result.

We compared the success rate of automatic matching with naive approaches and the proposed method. We performed automatic matching using 3-D shape data that have error with respect to the ground truth, and compared the success rate of the automatic matching in the case of using the proposed method and naive approaches. In this experiment, we used data taken around the fracture surface of the stone 1 and the stone 2 as shown in Fig. 6 in order to reduce the time required for one attempt and to perform a number of attempts. We tried the above process 3,500 times with each method.

Fig. 4 3-D data of quarried stone. **a** Stone 1. **b** Stone 2

(a)   **(b)**

Fig. 5 Automatic matching results by the proposed method. **a** Initial position. **b** After matching

(a)

(b)

Fig. 6 Range of data used in the experiment (*gray* area)

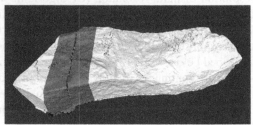

We compared three types of methods: (A) ICP algorithm, (B) multi-resolution ICP algorithm and (C) the proposed method. (A) uses ICP algorithm that is a part of the proposed method, and uses the same data as those used in the third stage of (B). (B) uses only the multi-resolution ICP algorithm.

4.1 Experimental Procedure

Automatic matching was performed in each method from a given state, which was translated and rotated with respect to the ground truth. The result was determined as success when the error between the position obtained by means of automatically matching and the ground truth was within the threshold described later. The ground truth was the position and orientation minimize the sum of the Euclidean distance between 20 sets of corresponding pairs given manually.

Automatic matching result is evaluated by the error of the estimated position and orientation T' obtained in the automatic matching and the ground truth T. Rotation angle θ given to the ground truth can be calculated by the following equation formula Rodriguez.

$$\theta = \cos^{-1}\left(\frac{\text{trace}(\boldsymbol{R}_{T'}) - 1}{2}\right) \tag{11}$$

The result of automatic matching was defined as success when translation error d_e was less than 45 mm and rotation error θ_e was less than 10°. Error of translation with respect to the ground truth d_e was represented by $d_e = ||\boldsymbol{t}_M - \boldsymbol{t}_{T'}||$. Rotational error θ_e was obtained by applying Eq. (11) the matrix $\boldsymbol{R}_{T'}\boldsymbol{R}_T^T$ obtained by multiplying the inverse rotation of $\boldsymbol{R}_T$ and the rotation $\boldsymbol{R}_{T'}$. $\boldsymbol{R}_T$ was rotation parameters of the position and orientation T, $\boldsymbol{t}_T$ was positional parameters.

4.2 Result

Figure 7 shows the graph of the success rate and the value of θ, Fig. 8 shows the graph of the success rate and the value of the distance given the translational movement with respect to the ground truth d ($d = ||\boldsymbol{t}||$). The success rate of (B) is higher than (A).

Fig. 7 Success rate of automatic matching with respect to the initial posture

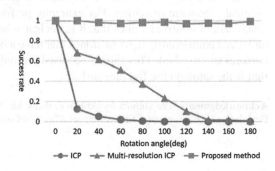

Fig. 8 Success rate of automatic matching with respect to the initial position

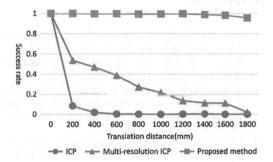

This is because (B) is less likely to converge to a local optimal solution as compared to (A), since multi-resolution ICP algorithm is possible to perform coarse registration using data of low resolution. With increasing the value of d and θ, the success rate of both (A) and (B) have fallen significantly. This is because the performance of the ICP algorithm depends on the shape and it is hard to align shape whose ratio of a characteristic portion is small compared with the size of the whole such as stone fracture surface. Fortunately, success rate of (C) was approximately 1.0 irrespective of the initial position. It is believed that there is a need to perform an initial position estimate with good accuracy in addition to the multi-resolution ICP algorithm from the above results.

5 Conclusion

In this study, we have proposed a dedicated method that performs automatic matching by using the plug-and-feather holes. In the experiments, it was confirmed that automatic matching of stones can be performed by using the plug-and-feather holes from the experiment using the 3-D shape data obtained by measuring the quarried stones. Further, the proposed method that made the initial position estimation using the plug-and-feather showed better results as compared with the case of using only the ICP algorithm.

Our future plan is to establish a suitable evaluation method for practical archeological automatic matching. For example, performing automatic matching using much data of quarried stones that it is unclear whether they have been a pair originally, and enumerating pairs of stones that have a high probability that their fracture surfaces are matched. This can also be used for making a visualization of the portion that is the same on the fracture surface.

Acknowledgments The authors would like to thank Mr. Y. Mikame of Kanagawa Archaeology Foundation for his supports of measurement of quarried stone.

References

1. Arun, K.S., Huang, T.S., Blostein, S.D.: Least-squares fitting of two 3-D point sets. IEEE Trans. Pattern Anal. Mach. Intell. **9**(5), 698–700 (1987)
2. Besl, P.J., McKay, N.D.: Method for registration of 3-D shapes. IEEE Trans. Pattern Anal. Mach. Intell. **14**(2), 239–256 (1992)
3. Brown, J.B., Toler-Franklin, C., Nehab, D., Burns, M., Dobkin, D., Vlachopoulos, A., Doumas, C.: A system for high-volume acquisition and matching of fresco fragments: reassembling Theran wall paintings. In: Proceedings ACM SIGGRAPH 2008. vol. 27, Issue 3, Article No. 84 (2008)
4. Chen, Y., Medioni, G.: Object modelling by registration of multiple range images. In: Proceedings of IEEE International Conference on Image and Vision Computing (ICIVC), vol. 3, pp. 2724–2729 (1991)

5. Chum, O., Matas, J.: Matching with PROSAC-progressive sample consensus. In: Proceedings of IEEE Computer Society Conference on Computer Vision and Pattern Recognition (CVPR 2005), vol. 1, pp. 220–226 (2005)
6. Huang, Q.-X., Flory, S., Gelfand, N., Hofer, M., Pottmann, H.: Reassembling fractured objects by geometric matching. Proc. ACM SIGGRAPH **25**(3), 569–578 (2006)
7. Ikeuchi, K., Miyazaki, D.: Digitally Archiving Cultural Objects. Springer, New York (2007)
8. Ikeda, S., Morita, K., Yoshimoto, S., Iwai, D., Sato, K.: SENS-Wiper: a wearable 3-D measurement system using kinect. Proc. Jpn. Soc. Archaeol. Inf. **11**, 75–80 (2013). (in Japanese)
9. Johnson, A.E., Martial, H.: Using spin images for efficient object recognition in cluttered 3D scenes. IEEE Trans. Pattern Anal. Mach. Intell. **21**(5), 433–449 (1999)
10. Jost, T., Hugli, H.: A multi-resolution scheme ICP algorithm for fast shape registration. In: Proceedings of First International Symposium on 3D Data Processing Visualization and Transmission, pp. 540–543 (2002)
11. Klein, G., Murray, D.: Parallel tracking and mapping for small AR workspaces. In: Proceedings of 6th IEEE and ACM International Symposium on Mixed and Augmented Reality (ISMAR 2007), pp. 1–10 (2007)
12. Koller, D., Trimble, J., Najbjerg, T., Levoy, N.G.: Fragments of the city: Stanford's digital forma Urbis Romae project. In: Proceedings of Third Williams Symposium on Classical Architecture, Journal of Roman Archaeology Suppl. vol. 61, pp. 237–252 (2006)
13. Levin, D.: Mesh-independent surface interpolation. In: Brunnett, G., Haman, B., Mueller, H. (eds.) Geometric Modeling for Scientific Visualization, pp. 37–49. Springer, Berlin (2004)
14. Masuda, T.: Registration and integration of multiple range images by matching signed distance fields for object shape modeling. Comput. Vis. Image Underst. **87**(1), 51–65 (2002)
15. Newcombe, R.A., Izadi, S., Hilliges, O., Olyneaux, D., Kim, M.D., Davison, A.J., Kohli, P., Shotton, J., Hodges, S., Fitzgibbon, A.: KinectFusion: Real-time dense surface mapping and tracking. In: Proceedings of 10th IEEE International Symposium on Mixed and Augmented Reality (ISMAR), pp. 127–136 (2011)
16. Rabbani, T., van den Heuvel, F., Vosselmann, G.: Segmentation of point clouds using smoothness constraint. Int. Arch. Photogram. Remote Sens. Spat. Inf. Sci. **36**(5), 248–253 (2006)
17. Rusinkiewicz, S., Levoy, M.: Efficient variants of the ICP algorithm. In: Proceedings of 3rd International Conference on 3-D Digital Imaging and Modeling, pp. 145–152 (2001)
18. Steder, B., Rusu, R.B., Konolige, K., Burgard, W.: Point feature extraction on 3D range scans taking into account object boundaries. In: Proceedings of the IEEE International Conference on Robotics and Automation (ICRA2011), pp. 2601–2608 (2011)
19. Winkelbach, S., Westphal, R., Goesling, T.: Pose estimation of cylindrical fragments for semi-automatic bone fracture reduction. In: Proceedings of 25th Annual Symposium of the German Association for Pattern Recognition (DAGM 2003), vol. 2781, pp. 566–573 (2003)
20. Zhang, Z.: Iterative point matching for registration of free-form curves and surfaces. Int. J. Comput. Vision **13**(2), 119–152 (1994)

A Route Recommender System Based on the User's Visit Duration at Sightseeing Locations

Daisuke Kitayama, Keisuke Ozu, Shinsuke Nakajima
and Kazutoshi Sumiya

Abstract The time spent at a sightseeing location differs for each tourist. If a tourist is particularly interested in a location, the visit duration may be long. In this case, we consider that the tourist's visit duration is longer than that of a typical tourist at similar sites. We assume that this difference shows the tourist's characteristics, such as the type of objects in which s/he is interested and his or her walking speed. It also indicates the level of congestion in the sightseeing area. Therefore, we propose a route recommender system whose recommendations are based on the difference between a user's and the typical visit duration at sightseeing locations. We focus on the category structure of the locations and the difference between the tourist's and the typical visit duration for estimating the visit duration at other locations. In this study, we developed a prototype system and evaluated it using a few sightseeing locations.

Keywords Recommender system · Sightseeing · Difference amplification · Visit duration

1 Introduction

Recently, the use of location-based systems, such as car navigation systems and smart phones, has become widespread. Accordingly, the number of location-based services that use the user's current location and movement history has increased.

D. Kitayama (✉)
Faculty of Information Science and Engineering, Kogakuin University,
Shinjuku, Tokyo, Japan
e-mail: kitayama@cc.kogakuin.ac.jp

K. Ozu
Meitec Corporation, Fujisawa, Kanagawa, Japan
e-mail: keisukeozu@gmail.com

S. Nakajima
Faculty of Computer Science and Engineering, Kyoto Sangyo University, Kyoto, Japan
e-mail: nakajima@cse.kyoto-su.ac.jp

K. Sumiya
Faculty of Human Science and Environment, University of Hyogo, Kobe, Hyogo, Japan
e-mail: sumiya@shse.u-hyogo.ac.jp

© Springer International Publishing Switzerland 2015
R. Lee (ed.), *Software Engineering Research, Management and Applications*,
Studies in Computational Intelligence 578, DOI 10.1007/978-3-319-11265-7_14

For example, smart phones are equipped with a navigation system not only for drivers in cars, but also for pedestrians. Therefore, tourists use both guidebooks and smart phones equipped with a GPS function. Most tourists use Web search, maps, and navigation applications. When users plan a sightseeing route, they decide to visit certain locations and determine the time they will spend there. However, planned visit durations are often changed during the trip, because the tourists discover interesting objects, their walking speed is slow, or a sightseeing area is congested. Then, their visit durations at other locations are also changed. It is obvious that these changes are important, because they sometimes restrict the time that the user can spend at other destinations. Recently developed car navigation systems have many functions, such as recommending a detour based on traffic congestion and frequently visited tourist locations. Therefore, these navigation functions have become a convenience for the user. However, the conventional navigation system does not recommend other tourist routes and locations when a tourist changes his or her visit duration at a particular location.

We consider that the various reasons why a tourist stays at a location for a long time, for example, s/he has an interest in that type of site or likes sightseeing slowly, constitute tourist characteristics. We consider that, if the reason for changing a visit duration is based on the tourist's level of interest, we can estimate that the duration of visits at the same type of tourist location will change accordingly. Conversely, if the reason for changing the duration of a visit is based on the tourist's characteristics, we can estimate that visit durations at all the sites will change. Therefore, we define the relationship between sightseeing locations, and calculate the estimated visit duration at each location using the relationship between sightseeing locations and the difference between the tourist's and the typical visit durations. Then, the visit durations at the same types of locations are estimated based on the difference between the typical and the user's visit durations, as representing the effect of the user's interests. Conversely, the visit durations at different types of locations are estimated based on reducing the difference between the visit durations, as representing the effect of tourist characteristics. Our system then recommends a new tourist route using the estimated visit duration.

2 Our Approach

2.1 Concept of Our Route Recommender System

Recent navigation systems are accurate and users find them convenient. However, these systems are designed such that they show only the shortest route and therefore they are not suitable for a trip during which the user frequently changes her or his plans according to the time s/he spends at locations. We consider that there are two reasons for changing the visit duration. The first is the user's interest in the type of location. For example, if a user likes visiting temples, s/he may stay at temples for

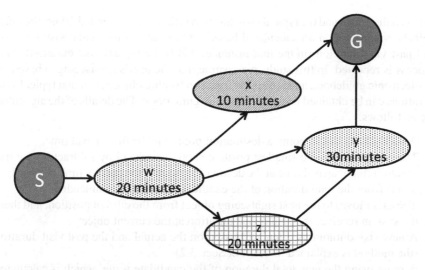

Fig. 1 Model of sightseeing objects and visit durations

a long time. The second reason is the user's/sighseeing area's characteristics. For example, if a user walks slowly or the sightseeing area is congested, the user may stay at all locations for a long time. Figure 1 shows a model of sightseeing objects and visit durations. There are four objects between the origin point and the destination point. The color of the object shows its type. Therefore, if objects have the same color, they are of the same type.

We now explain the concept of our route recommendation system. First, the recommended route is "$S - w - x - G$". We assume that a user stayed at sightseeing object w 10 min less than the typical visit duration. In this case, the system may recommend route "$S - w - y - G$", because the user has time for additional sightseeing. In addition, it is considered that the user can perform the amount of sightseeing that satisfies her or him in a short time. Therefore, the system should recommend route "$S - w - z - y - G$".

To realize the above concept, we define the relationship between sightseeing objects and we calculate the estimated visit duration at each object using the relationship between the sightseeing objects and the difference between the actual visit duration of the user and the typical visit duration. Then, the visit durations at the same types of objects are estimated based on the difference between the typical and the tourist visit durations, as the effect of the user's interests. However, the visit duration at different types of objects is estimated based on reducing the difference between visit durations, as the effect of the user's characteristics. The system recommends a new tourist route using the estimated visit duration.

We now describe the procedure of our method. First, a user sets an origin point and a destination point, and then, our system recommends a route based on tourists' typical visit durations. Next, the user performs the sightseeing, and then, his/her actual visit duration at a sightseeing object is extracted. A new route based on the difference between the actual and the typical visit duration is then recommended. Then, the other objects' visit duration are calculated based on the ratio of the actual visit duration and past visit duration (in the first instance, this is the typical visit duration). This process is repeated. In this study, we assume that the user's sightseeing is based on an electronic guidebook and information on sightseeing objects, and that typical visit durations can be obtained from the electronic guidebook. The details of the algorithm are as follows.

1. A user sets an origin point, a destination point, and his/her arrival time.
2. The route that has the smallest positive score calculated by subtracting the trip duration (input arrival time at the destination point—the current time at the origin point) from the total duration of the candidate route is recommended.
3. The user moves to the next sightseeing object from the current position, and then, the system receives the actual visit duration at the current object.
4. A new visit duration is calculated based on the actual and the past visit duration (the method is explained in detail in Sect. 3.2).
5. A route using the new total duration of the candidate route, which is calculated using the new visit duration, is recommended.
6. Steps 3 to 5 are repeated while some of the time allotted to the trip remains and there are still visitable sightseeing objects.
7. If all of the candidate routes have a negative score, a direct route to the destination point from the current object is recommended.

2.2 Related Work

We now introduce conventional studies on personal navigation systems and the relationship between geographical objects. Conventional personal navigation systems are widely available. Baus et al. [2] proposed a dynamic guiding route system based on the user's context (e.g., location, time, and environment). Butz et al. [3] developed indoor navigation systems. Then, Cheverst et al. [4, 5] proposed a system for providing sightseeing information to mobile devices. Rehrl et al. [11] developed an integrated system providing travel planning, information on transportation, and indoor/outdoor navigation. Basic techniques, such as Dijkstra's algorithm [6] and the A* algorithm [8], exist for obtaining the shortest path between two positions. These techniques are widely used in car navigation systems. We use all combinations of objects for generating route recommendations in this study, because we focus only on calculating the estimated visit duration. Certainly, we can use these techniques for optimizing computational complexity. In this part of the method, we may also use a generic algorithm, predatory search, and so on.

Nakajima et al. [7, 10] proposed a route recommendation method using the difference in the features of a recommended and a user-selected route, based on a difference amplification algorithm for car navigation systems. They assume that the reason for the user's route selection, when s/he changes her or his driving route is indicated by the difference in the features of the recommended and the user-selected route. For example, a car navigation system recommends the shortest route; however, the user wants to drive on a wide road. In this case, the user changes the driving route in order to use a wider road than that included in the recommended route. A conventional car navigation system may re-recommend the shortest route. However, the proposed system generates a new recommended route with a wide road. We use this difference amplification approach for estimating the visit duration.

Studies have been conducted on the relationship between geographical objects. Arikawa et al. [1] proposed a method for detecting visible objects using the ontology of geographical objects for adaptation to a user's needs. Shimada et al. [12] and Inoue et al. [9] developed a method for selecting objects using attributes, such as type and position. Arikawa et al. used the category of objects as their attribute. Shimada et al. used their positional relationship and direction as attributes of objects. Inoue et al. used the visibility of objects as their attributes. In this study, we use the hierarchical category as the attribute of objects.

2.3 Preliminary Experiment

We conducted a preliminary experiment for confirming the relationship between visit durations and the types of objects. This preliminary experiment was based on a questionnaire. We used 11 objects (5 temples, 3 museums, 3 parks), their typical visit duration, and their abstracted description from a guidebook. The participants were seven university students. They were asked to estimate how much time they would spend visiting these sightseeing objects. The experimental results are shown in Table 1. The results for most participants differed from the typical visit duration. We explain the results in detail as follows. Some participants estimated a visit duration of 0 min. This duration means that the participant has no interest in this object, and therefore, s/he would not visit it.

We calculated the correlation coefficient between the participant's estimated visit duration and each object. We used a correlation coefficient of 0.4 as the threshold, because this value means that the strength of the relationship is greater than the median. The number of all pairs of objects is 55. The correlation coefficients for 15 pairs of objects (27 %) are greater than the threshold. The correlation coefficients for 4 pairs of temples, 2 pairs of museums, and 2 pairs of parks are greater than the 0.4 threshold. Therefore, we confirmed that the visit durations are related to the type of object.

Table 1 Answers of participants and typical visit duration (the unit is minutes)

	Typical visit duration	A	B	C	D	E	F	G
Kaizouji Temple	20	20	20	20	30	20	20	40
Tokeiji Temple	30	30	40	30	40	20	30	40
Jouchiji Temple	30	30	0	30	40	20	20	20
Jyufukuji Temple	15	0	10	15	20	20	20	30
Tsurugaoka Hachimangu Shrine	60	60	90	40	40	40	40	40
Kamakura Museum of Literature	40	0	60	30	40	30	20	30
Kamakura Museum of National Treasures	60	60	90	50	50	60	50	50
The Museum of Modern Art, Kamakura Annex	60	60	60	40	50	60	40	40
Genjiyama Park	30	15	0	30	40	40	20	20
Kewaizaka Park	10	30	0	5	5	10	15	10
Shishimai Park	30	30	0	20	30	40	30	40

3 Route Recommendation Using the Difference Between the Typical and the User's Visit Duration

3.1 User's Intention and Sightseeing Objects

We consider that a user's visit duration at a location is changed according to two intentions. The first is based on the user's interest: the duration of the visit becomes longer when the user is interested in the objects. The second intention is based on the user's characteristics: the length of the visit becomes longer when the user walks slowly. We change the calculation parameter for estimating the visit duration based on the type of user's intention. We consider that the user's interests have a strong influence when the objects are of the same type and that the user's characteristics have a weak influence on the visit duration of all objects.

We now explain the definition of sightseeing objects and paths. A sightseeing object has a visit duration value (the default value is the typical visit duration) and a type of object. A type of object is modeled by a DAG (directed acyclic graph) structure. For example, a sightseeing object as a type of object is the root level. The children of a sightseeing object are artificial objects and natural objects. The children of artificial objects are temple, shrine, and museum, and of natural objects are temple, shrine, and park. In this study, we used this DAG structure (see Fig. 2). We assume that the DAG structure of the type of object can be prepared according to an electronic guidebook. Another method for obtaining a DAG structure is to cluster using the feature vectors of objects. The path is a road between two sightseeing objects. The path has a moving time. We use a graph that consists of sightseeing objects as nodes and paths as edges.

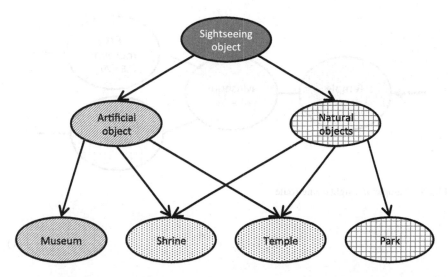

Fig. 2 DAG structure of type of object

3.2 Calculation of Estimated Visit Duration

The proposed system recommends a route by reflecting the difference between an object's actual visit duration and past visit duration onto the visit duration of other objects. The actual visit duration time when the user leaves a sightseeing object is extracted. Figure 3 shows an example of a sightseeing route. In this figure, the number is the typical visit duration plus the moving time (the unit is minutes) and the color shows the type of object, where grided represents natural objects, diagonal represents artificial objects, and doted represents objects that are of both types.

We now explain how to reflect the difference in the visit duration onto other objects' visit duration. The ratio of the actual and the past visit duration is used. Then, the denominator is the past visit duration and the numerator is the actual visit duration. In other words, when the actual visit duration is greater than the past visit duration, the visit duration of other objects should be amplified. The ratio is adjusted using the amplification parameter by power calculation. In addition, if another object is the same type of user-visited object, this ratio is used for amplifying the size. The amplification parameter is set at a value greater than 1.0 when a target object is the same type of user-visited object, because the object is influenced by the user's interest and the user's characteristics. However, if another object is of a different type of user-visited object, this ratio of reduced size is used. The amplification parameter value is set at less than 1.0 when a target object is of a different type of user-visited object, because a different object is influenced only by the user's characteristics. We calculate the estimated visit duration by

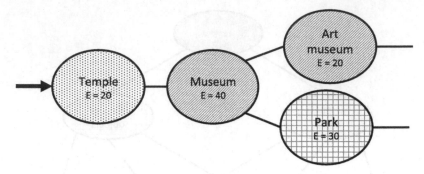

Fig. 3 Example of a sightseeing route

$$E'_y = E_y \times \left(\frac{A_x}{E_x}\right)^n \tag{1}$$

E'_y New estimated visit duration of target object y
E_y Past estimated visit duration of target object y
E_x Estimated visit duration of user-visited object x
A_x Actual visit duration of user-visited object x
 n The amplification parameter.

The amplification parameter is determined based on the distance between the type of target object and on the type of user-visited object in the DAG structure. In this study, we set the amplification parameter as 1.0 when the target object was the same type of user-visited object and as 0.6 when the target object was a different type of user-visited object. If two objects had the same parent on the DAG structure, we determined that these objects were of the same type and if they had different parents we determined that the objects were of a different type.

We show examples of the calculation in Figs. 4 and 5. Figure 4 shows that a user visited a temple for 15 min. The past visit duration of this temple is 20 min. Museum, art museum, and park have the same parent as the temple in the DAG structure. Therefore, the estimated visit durations of museum, park, and art museum is calculated using a ratio 15/20 and the amplification parameter 1.0. In other words, it is estimated that the visit durations of this user for museum, park, and art museum are equally short. Figure 5 shows that, after visiting the temple, the user visited a museum for 40 min. The past visit duration time of the museum is 30 min. Art museum is of the same type as museum. Therefore, the estimated visit duration of the art museum is calculated using a ratio 40/30 and the amplification parameter 1.0. On the other hand, the park is not of the same type as the museum. Therefore, the estimated visit duration of the park is calculated using a ratio 40/30 and the amplification parameter 0.6. Thus, it can be estimated that this user stays at artificial objects for a long time and at natural objects for a short time.

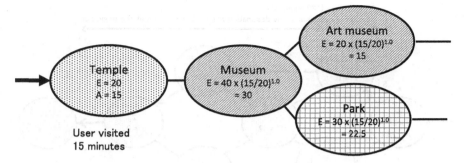

Fig. 4 Calculation result when the user visited a temple for 15 min

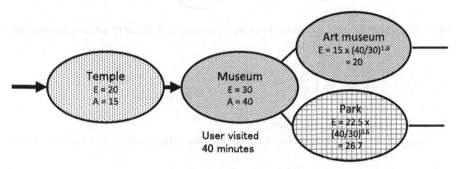

Fig. 5 Calculation result when the user visited a museum for 40 min

3.3 Route Recommendation Using Estimated Visit Duration

We now describe our proposed route recommendation method based on the estimated visit duration. Our system recommends the route that has the smallest positive score calculated by subtracting the trip time (input arrival time at the destination point—current time at the origin point) from the total time of the candidate route. In this case, we may use a genetic algorithm for route recommendations; however, we will address this point in future studies.

First, the actual visit duration of a user-visited object is extracted. The estimated visit duration of all other objects is calculated using the formula 1. The total time of all candidate routes is calculated. In this case, we can use a genetic algorithm to generate the route recommendation; however, in this study, the candidate route is simply extracted by using all the combinations. Then, we obtain the moving time from an object to another object using the Google Maps API. Finally, the route that has the smallest positive score is recommended. The score for the recommendation is calculated using

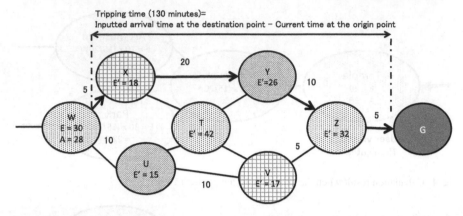

Fig. 6 Example of recommended route when user leaves object W. E and E' are the estimated visit durations. A is an actual visit duration

$$Score_i = \begin{cases} Time_{trip} - Time_{C_i} & (Time_{trip} \geq Time_{C_i}) \\ 0 & (Time_{trip} < Time_{C_i}) \end{cases} \tag{2}$$

C_i The ith candidate route

$Time_{trip}$ Trip time (input arrival time at the destination point—current time at the origin point).

$Time_{C_i}$ Total time of candidate route C_i.

Figure 6 shows an example of a route recommendation. In this example, it is assumed that the user stays at object A for 28 min. In addition, the object's typical visit duration is 30 min. The estimated visit duration of other objects is calculated based on the difference between the actual and the typical visit duration and on the type of object. The current time at the origin point is 14:20 and the input arrival time at the destination point is 16:30. Therefore, the trip duration $Time_{trip}$ is 130 min. The score of candidate route "$W - X - Y - Z - G$" is 14 ($130 - 116$). In detail, the total time of candidate route "$W - X - Y - Z - G$" (116) is calculated by adding the total visit duration ($18+26+32$) to the total moving time ($5+20+10+5$). However, the score of another candidate route "$W - U - V - Z - G$" is 36 ($130 - 94$). In this case, the system should recommend route "$W - X - Y - Z - G$", because this route allows the user to stay at the sight seeing spot for a long time until the input arrival time at the destination point.

4 Evaluation

We conducted an experiment using a questionnaire to confirm the effectiveness of our method. This questionnaire was similar to that used in our preliminary experiment. The participants in this experiment were 11 university students. We asked the participants how long they would stay at eight sightseeing objects in Kamakura,

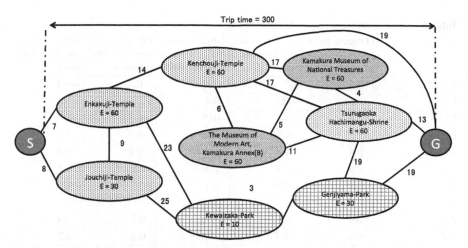

Fig. 7 Experimental data. E is an estimated visit duration. The number close to the edge is the moving time

Japan. The objects consisted of six artificial and six natural objects. In addition, four objects were of both the artificial and natural type. The participants decided how long they would spend at the sightseeing objects according to descriptions of the sightseeing spot, typical visit duration, and a map indicating the location of each of the sightseeing spots (Fig. 7). Then, the participants gave the reason for their decision.

We calculated the estimated visit duration using three objects as user-visited objects. Our selection of objects for this experiment was based on the following five patterns.

- **Objects are of the same type**: We calculated the estimated visit duration using three temples.
- **Objects are of different types**: We calculated the estimated visit duration using one artificial object, one natural object, and one object that belongs to both types.
- **Objects have the longest visit duration**: We calculated the estimated visit duration using the top three objects ordered from the longest to the shortest visit duration.
- **Objects have various visit durations**: We calculated the estimated visit duration using the second, fifth, and seventh objects ordered from the longest to the shortest visit duration.
- **Objects have the shortest visit duration**: We calculated the estimated visit duration using the top three objects ordered from the shortest to the longest visit duration.

We confirmed the influence of the type of object on the length of the visit.

We calculated the correlation coefficient between two data: the difference between the estimated visit duration using our method and the typical visit duration, and the difference between the participant's input visit duration and the typical visit duration.

Table 2 Experimental results: correlation coefficients

	Objects are of the same type	Objects are of difference types	Objects have the longest visit duration	Objects have various visit duration	Objects have shortest visit duration
P1	0.5	–	−0.7	−0.1	−0.7
P2	0.5	−0.1	0.4	–	–
P3	0.8	0.4	−0.5	−0.3	−0.8
P4	−0.6	0.3	−0.4	−0.3	−1.0
P5	0.8	−0.6	−0.3	0.4	0.5
P6	0.9	−0.8	0.4	0.8	0.7
P7	0.8	−0.9	0.1	0.7	0.5
P8	0.1	0	−0.4	–	–
P9	0.9	−0.6	−0.4	0.7	0.4
P10	0.9	−0.7	−0.8	0.3	0.2
P11	0.9	−0.5	0.5	0.4	−0.2
Correlation coefficient ≥ 0.7	7	0	0	3	1
Correlation coefficient ≥ 0.4	9	1	3	5	4

We show the experimental results in Table 2. The columns headed P_1, P_2, ... , P_{11} show the correlation coefficient for a participant. The other two columns show the number of participants for whom the correlation coefficient was ≥ 0.7 or ≥ 0.4. - denotes that the correlation coefficient could not be calculated because the difference between the estimated visit duration using our method and the typical visit duration was 0.

First, we explain the influence of the type of visited object. For the case "objects are of the same type", the correlation coefficient is high, whereas for the case "objects are of different types" the correlation coefficient is low. We supposed that our method would be able to adjust the estimated visit duration for objects of different types. In our opinion, the performance of our method would be improved by adding to the database visited objects that are of the same type as these cases. In these cases, not only the object type condition but also the visit duration condition are different. Objects have various visit durations in the case of "objects have the same type". Conversely, objects have a biased visit duration in the case of "objects are of different types".

Now, we explain the influence of the length of the visit duration. For the case "objects have various visit durations", the correlation coefficient is high. On the other hand, for the cases "objects have the longest visit duration" and "objects have the shortest visit duration" the correlation coefficients are low. It can be said that the system can estimate the visit duration when the visited objects have various visit durations. In this case, we can obtain the characteristics of the user's sightseeing visit. When our method receives only a long visit duration, it estimates a visit duration that is longer than the typical one for all objects, and if it receives only short visit durations, it estimates a visit duration that is shorter. In addition, the objects include various types of objects in the case of "objects have various lengths of visit duration".

According to the above results, our method can calculate the estimated visit duration using the difference between the typical and the tourist's visit duration and the object type. In particular, the estimate of the visit duration was accurate when the visited objects had various actual visit durations. In future work, we should evaluate our system using a large scale data set that also includes other locations. In addition, we should perform experiments using not only a questionnaire but also an actual trip.

5 Conclusion

In this paper, we proposed a route recommendation method based on the difference between the actual and the estimated visit duration of a visitor at a sightseeing location. In particular, we changed the influence of this difference for updating the estimated visit duration according to the type of object. Thus, we could calculate the estimated visit duration that is suited to the user's sightseeing characteristics. In this study, we conducted an experiment in which participants responded to items of a questionnaire regarding how long they would stay at certain tourist locations. We used five patterns of the tendency of the assumed user's trajectories. We calculated the correlation coefficient between the participant's answers and the estimated visit duration. The results of this experiment confirmed that our method estimates the visit duration accurately when the objects that the users visited have various visit durations.

In our future work, we intend to evaluate our method using actual sightseeing behavior. For instance, we will confirm the satisfaction of the users with the recommended route as compared to a route that is planned without using our system. Then, we will extend the type of object. We used only two types of object, artificial and natural, in this study. In our opinion, we should subdivide the types of object. In addition, we should define the degree of interest in an object using the numbers of tourists who take photos and send micro blogs.

Acknowledgments This work was supported by JSPS KAKENHI Grant Numbers 24700098, 26280042.

References

1. Arikawa, M., Kambayashi, Y.: Dynamic name placement functions for interactive map systems. Aust. Comput. J. **23**(4), 133–147 (1991)
2. Baus, J., Kruger, A., Wahlster, W.: A resource adaptive mobile navigation system. In: Proceedings of 2002 International Conferences on Intelligent User Interfaces 2002 (IUI-02), pp. 15–22 (2002)

3. Butz, A., Baus, J., Kruger, A., Lohse, M.: A hybrid indoor navigation system. In: Proceedings of 2001 International Conferences on Intelligent User Interfaces 2001 (IUI2001), pp. 25–33 (2001)
4. Cheverst, K., Davies, N., Mitchell, K., Friday, A.: The design of an object model for a context-sensitive tourist guide. Comput. Graph. J. **23**(883–891), 883–891 (1999)
5. Cheverst, K., Davies, N., Mitchell, K., Friday, A., Efstratiou, C.: Developing a context-aware electronic tourist guide: some issues and experiences. In: Proceedings of 2000 ACM Special Interest Group on Computer-Human Interaction (SIGCHI- 00), pp. 17–24 (2000)
6. Dijkstra, E.W.: A note on two problems in connexion with graphs. Numer. Math. **1**(1), 269–271 (1959)
7. Hamada, K., Nakajima, S., Kitayama, D., Sumiya, K.: Route recommendation method based on driver's intention estimation considering the route selection when using the car navigation. In: Proceedings of the 2014 IAENG International Conference on Data Mining and Applications, IMECS2014, pp. 383–388 (2014)
8. Hart, P., Nilsson, N.: Raphael, B.: A formal basis for the heuristic determination of minimum cost paths. IEEE Trans. Syst. Sci. Cybern. **4**(2), 100–107 (1968)
9. Inoue, T., Nakazawa, K., Yamamoto, Y., Shigeno, H., Okada, K.: Use of human geographic recognition to reduce GPS error in mobile mapmaking learning. In: Proceedings of Fifth International Conference on Networking and the International Conference on Systems (ICN / ICONS / MCL 2006), p. 222 (2006)
10. Nakajima, S., Kitayama, D., Sushita, Y., Sumiya, K., Chandrasiri, N.P., Nawa, K.: Route recommendation method for car navigation system based on estimation of driver' s intent. In: Proceedings of the 2012 IEEE International Conference on Vehicular Electronics and Safety (ICVES 2012), pp. 318–323 (2012)
11. Rehrl, K., Leitinger, S., Bruntsch, S., Mentz, H.J.: Assisting orientation and guidance for multimodal travelers in situations of modal change. In: Proceedings of 2005 IEEE International Conferences on Intelligent Transportation Systems (ITSC-05), pp. 407–412 (2005)
12. Shimada, S., Tanizaki, M., Maruyama, K.: Ubiquitous spatial-information services using cell phones. IEEE Micro **22**(6), 25–34 (2002)

Predicting Access to Healthcare Using Data Mining Techniques

Sergey Shishlenin and Gongzhu Hu

Abstract Healthcare is a very basic need for everyone in today's society. However, many individuals do not have or have difficulties accessing healthcare services. Researchers have studied various aspects in healthcare, including using data mining techniques to analyze healthcare data. Most of these studies, however, focused on treatment effectiveness, healthcare management, fraud and abuse detection, etc., few have reported on the accessibility study of healthcare. In this paper, we examine individual demand of receiving health care using the data from the Behavioral Risk Factor Surveillance System, a sample of the U.S. population. Models were built using the typical predictive modeling methods to predict the accessibility of healthcare based on the explanatory variables in the data set. Our experimental results showed that the regression and neural network models yielded better prediction accuracy whereas the decision tree and nearest-neighbor models fell behind in their performance.

Keywords Healthcare access · Data mining · Predictive modeling · Classification

1 Introduction

Healthcare is a complex problem involving many fields of study, such as medicine, management, economics, social study, public policy, statistics, and information technology. At the core of healthcare research is the analysis of data. Data mining, a term referring to data analysis for the purpose of discovering useful information/knowledge from the data, has been used extensively for many organizations and "is becoming increasingly popular, if not increasingly essential" in healthcare [11].

In healthcare data mining research, most work has been done in the areas of treatment effectiveness, healthcare management, Customer relationship management, and

S. Shishlenin · G. Hu (✉)
Department of Computer Science, Central Michigan University,
Mt. Pleasant, MI 48859, USA
e-mail: hu1g@cmich.edu

S. Shishlenin
e-mail: shish1s@cmich.edu

© Springer International Publishing Switzerland 2015

191

R. Lee (ed.), *Software Engineering Research, Management and Applications*,
Studies in Computational Intelligence 578, DOI 10.1007/978-3-319-11265-7_15

fraud and abuse detection [11]. The issue of access to healthcare, however, has not been studied extensively using data mining approaches. There are many variables involved in healthcare access that are frequently included in widespread surveys. data mining provides an opportunity to develop models which may not only provide the best means to classify and predict individuals with poor access to health care, but also identify contributing factors which have not yet been identified. In this paper, we attempt to utilize data from the Behavioral Risk Factor Surveillance System [4] in pursuit of these ends.

This paper focuses on the analytical techniques to analyze individual access to health care using the SAS software, including base SAS 9.3 (for data merging and recoding purposes) and SAS Enterprise Miner (EM) 7.1 [14] (for data analysis). An introduction of SAS EM for predictive modeling can be found in [13].

2 Access to Health Care

The present study seeks to investigate some of these potential predictors of healthcare access by utilizing the data from the Behavioral Risk Factor Surveillance System (Centers for Disease Control and Prevention) that is "an ongoing data collection program designed to measure behavioral risk factors for the adult population" related to health practices [4].

Behavioral models have been developed for the use of health services in the health-care research community to capture the primary factors regarding the use health services by individuals and families. These models had evolved through several phases and an emerging model was formed in the mid 1990s [2] that includes multiple influences on health services' use and on health status. The model also includes feedback loops showing that the outcome would in turn affect subsequent predisposing factors and health behavior.

Hargraves and Hadley [8] examined and discussed how the coverage of health insurance can reduce disparities in access to care. They found that the lack of health insurance was the one of the most crucial factors in white versus Hispanic inequalities for all criteria and for two of the white versus African American disparities. Differences in income were ranked second in terms of importance. However, the positive effects of health insurance coverage in decreasing inequalities outweigh benefits of rising up physician charity care or emergency rooms access.

There are many factors (e.g., socioeconomic, dietary, alcohol consumption etc.) which affect individual's access to health care. Among factors, research suggests that an individuals' level of physical activity is a critical component in determining one's access via such indicators as disease risk reduction and improved mood (Centers for Disease Control and Prevention [5]). Physical activities refer to anything that adds beneficial movement to one's otherwise sedentary day.

Wiltshire et al [17] has objective of unraveling the relationships between race/ethnicity, socioeconomic status (SES) and inappropriate medical care needs.

They examined associations between inappropriate medical needs and SES among African American and White women using the data from the 2003–2004 Community Tracking Study Household Survey.

3 Predictive Modeling

Predictive modeling [12, 18] is a statistical method to estimate (predict) the outcome of a new observation (a data record) based on the knowledge obtained from a previous observations. Intuitively, this problem starts with an *information table* like the one shown in Table 1.

Formally, let D be the universal data set in an application domain, $X = \{x_1, \ldots, x_m\}$ be a set of m attributes (variables or predictors), and $C = \{c_1, \ldots, c_k\}$ be a set of k decision labels. Each observation (data record) $d_i \in D$ is a set of values in $X \times C$. Let $D_t = \{d_1, \ldots, d_n, d_i \in D\}$ be a set of data records with each d_i being associated with a known decision label c_j. Predictive modeling is to build a model (function f) from D_t, such that f is a mapping

$$f : X \to C$$

That is, given a data record $d = (b_1, \ldots, b_m) \in D$, the model should tell which decision label should be assign to d: $f(d) = c_j \in C$.

When applying predictive modeling to the access to healthcare, for example, the data set $D = \{person\}$ and the decision labels $C = \{has\ access, no\ access\}$, the task is to build a function f that applies to a person p: $f(p) = yes$ or no.

There are many different ways for building the model f. The very basic methods include regression and Bayesian classification. Other machine learning techniques have also been used for model building such as decision tree, neural networks, support vector machines (SVM), etc. These methods have been widely studied in the

Table 1 Predictive modeling

	Observation	Attributes X			Decision y
		x_1	......	x_m	
D_t	d_1	a_{11}	......	a_{1m}	c_1
	d_2	a_{21}	......	a_{2m}	c_2
	...	...	......	...	...
	d_n	a_{n1}	......	a_{nm}	c_n
New	d	b_1	......	b_m	?

literature, and comprehensive descriptions of these methods, among others, are given in the book [7].

In this study, we applied regression, decision tree, nearest neighbor, and neural networks to the healthcare data set.

4 Data Source

The data for our study was obtained from the Behavioral Risk Factor Surveillance System (BRFSS), which is a large-scale, national survey conducted annually. On average, the survey gathers health-related information from approximately 5,00,000 United States residents yearly. As the scope of our study is limited to those individuals living in the state of Michigan, only participants claiming that as their current state of residence were retained. The number of participants each year sampled from Michigan represents approximately 2.19 % of the overall sample, which is slightly under-sampled from the state's true proportion of the national population.

The number of participants is 504,408 in survey. The full sample for Michigan consists of 11,049 individuals over 400 variables.

As the number of variables present within the database is very large, data cleaning was a difficult task. We obtained the full database, removed participants not from the state of Michigan, and cleaned the data. The majority of data cleaning tasks consists of attempting to determine cases where the same questions were coded differently or given different titles, as well as handling the coding of missing values within the database. The coding schemes for items vary considerably within year and possibly between them, with different codes being used to denote true missing data (questions unasked), refusals to respond to questions, indications by the participants of not being sure of the correct answer, and actual responses provided by participants. Variables which ask for numerical data (i.e. how many days missed from work) are not only being retained in their default form so as to allow for differences between missing values, refusals, and being unsure of the answer, but also recoded with all of these being marked as missing data to allow for Enterprise Miner to take advantage of the interval or ratio nature of the data.

SAS program was written to recode these variables that were screened for inclusion in the database in the hope that decreasing the overall size of the database will make analysis more efficient. Variables that we thought to be not useful for the analysis (i.e. participant ID codes, whether the phone number dialed had the correct name attached to it, etc.) were dropped from the database prior to entry into Enterprise Miner.

Table 2 Descriptive statistics of health care access variable

Was there a time in the past 12 months when you needed to see a doctor but could not because of cost?	Freq	(%)
Yes	63,828	12.69
No	439,274	87.31
Missing	1,306	

4.1 Dependent Variable

The dependent variables in current analysis measure health care access. Table 2 shows that 63,828 of 439,274 respondents, or 12.69 %, has difficulties with getting health care.

4.2 Explanatory Variable

There are over 400 variables in the BRFSS survey data set. Here we only explain a few as shown in Table 3.

The statistics of some of these variables from the data set is given in Table 4.

Table 3 Variable descriptions

ID	Type	Description
CHECKUP1	Ordinal	How long has it been since you last visited doctor for routine checkup?
GENHLTH	Ordinal	General health of individual
PHYSHLTH	Ordinal	# of days during the last 30 days physical health was not good
CVDINFR4	Binary	Ever had heart attack (myocardial infarction)
CVDCRHD4	Binary	Ever had angina (coronary heart disease)
_ASTHMS1	Nominal	Asthma status
DIABETE3	Binary	Diabetes status
_RFHYPE5	Binary	High blood pressure
_SMOKER3	Nominal	Smoke status
EMPLOY	Nominal	Employment status
_LLCPM03	Ordinal	Education level
_INCOMG	Ordinal	Income category
SEX	Binary	Gender
_IMPRACE	Nominal	Race/ethnicity group
_BMI5CAT	Ordinal	Body mass index (BMI)

Table 4 Variable statistics

ID	Value	Freq	(%)
CHECKUP1	Within past year	360,620	73.18
	Within past 2 years	60,075	12.19
	Within past 5 years	36,083	7.32
	5 or more years ago	36,007	7.31
GENHLTH	Excellent	88,348	17.58
	Very good	160,222	31.89
	Good	155,227	30.90
	Fair	69,234	13.78
	Poor	29,379	5.85
CVDINFR4	Yes	29,983	5.97
	No	472,004	94.03
CVDCRHD4	Yes	3,0171	6.04
	No	469,638	93.96
_ASTHMS1	Current	45,087	9.00
	Former	18,075	3.61
	Never	437,957	87.40
_SMOKER3	Current (every day)	61,653	12.28
	Current (some days)	22,561	4.50
	Former smoker	147,542	29.40
	Never smoked	270,120	53.82
EMPLOY	Employed for wages	206,176	41.09
	Retired	141,841	28.27
	Self-employed	40,093	7.99
	Unable to work	36,335	7.24
	A homemaker	35,150	7.00
	Out of work ($\geq$1 year)	17,003	3.39
	Out of work (<1 year)	13,878	2.77
	A student	11,334	2.26
_INCOMG	<$15,000	55,052	10.91
	[$15,000, $25,000)	79,075	15.68
	[$25,000, $35,000)	51,507	10.21
	[$35,000, $50,000)	64,195	12.73
	$\geq$ $50,000	181,362	35.96
	Don't know/Not sure	73,217	14.52
SEX	Male	197,927	39.24
	Female	306,481	60.76

5 Analysis Methodology and Modeling

SAS Enterprise Miner 7.1 was used to analyze the healthcare data. In particular, several classification models were built and compared using SAS and the "best" model was selected. The SAS modeling diagram is shown in Fig. 1.

5.1 Missing Data Imputation

Missing data is a serious issue for building predictive models. Most of predictive modeling techniques delete the entire case for modeling if there is a missing data of an input. As a consequence, many cases may be deleted prior to modeling. The only technique that does not delete missing cases of input variables is the Decision Tree modeling.

SAS Enterprise Miner 7.1 is used to handle with a problem of missing data. A special node was added to the modeling schema (the Impute node in Fig. 1). It uses a special split technique for missing value imputation. The search for a split on input variables uses observations whose values are missing on the input. All such observations are assigned to the same branch. The branch may or may not contain other observations. The resulting branch maximizes the worth of the split. For splits on a categorical variable, this amounts to treating a missing value as a separate category. For numerical variables, it amounts to treating missing values as having the same unknown non-missing value.

One advantage of using missing data during the search is that the worth of split is computed with the same number of observations for each input. Another advantage is that an association of the missing values with the target values can contribute to the predictive ability of the split.

In order to make the best use of available data by replacing missing values

- Tree is used for categorical variables.
- Distribution is used for interval variables to preserve distributions.
- Single is used for indicator variables.

Another step is a date partition using the Data Partition node in SAS EM. It divides data into:

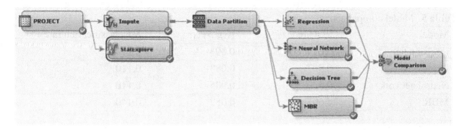

Fig. 1 Modeling scheme in SAS EM

- Training data (50 % of the data): for model building.
- Validation data (30 %): for validating the model obtained from the training data in order to select the "optimal" model.
- Testing data (20 %): a subset of the data held out that will be used for independent test of the model.

The following modeling methods were used in our study:

(1) **Regression**. It is the most simplified global model and it is parametric. Since the target variable is binary it is more reasonable to use logistic regression.
(2) **Decision Tree**. A tree is built based on the "splitting" variable at each level that has the most infromation gain.
(3) **K-Nearest Neighbor**. The kNN model classifies an observation based on the labels of its k nearest neighbors. The minimum bounding rectangle (MBR) approach was used in the model.
(4) **Neural Network**. The neural network framework encompasses various statistical models. The general form of a feed-forward neural network (with one hidden layer) expresses a transformation of the expected target as a linear combination of nonlinear functions of linear combinations of the inputs.

6 Results

SAS EM uses Model Comparison for selecting the final "best" model. The fit statistics between models were compared to choose the final model. The Model Comparison Node provides a full range of criteria for model comparison with results given in Table 5.

ROC (Receiver Operation Characteristic) curve is a chart for evaluating the performance of a model with the area underneath the curve as the measure of the "goodness" of the model—the larger the area, the better the model. The ROC curve of the models in our study is shown in Fig. 2. The baseline is the 50–50 guessing line with the area (0.5 value). A good model usually has the area over 0.8. It appears there is virtually very little difference between Regression and Neural Networks models while Decision Tree and MBR significantly falls behind for both validate and test data.

Table 5 Model comparison

Model	ROC index	Avg sq err	Misclassification rate
Decision tree	0.675	0.093	0.113
Regression	0.837	0.083	0.110
Neural network	0.829	0.085	0.110
MBR	0.603	0.082	0.140

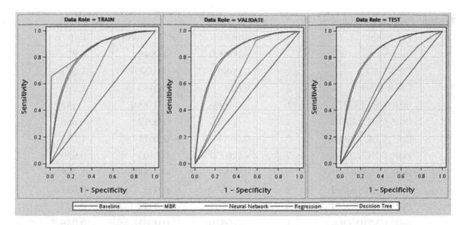

Fig. 2 ROC comparison plot

6.1 Regression

Logistric regression provides the most informative estimates based on comparison step. Moreover, it also indicates the significance of the 64 input variables. Table 6 provides some detailed information regarding the regression estimates and chi-square statistics as well as odds ratio. Parameters marked with a ∗ are the most significant, i.e. has a most effect on dependent variable.

There are a number of variables that greatly reduce the chance for individuals to have all their health care services. Some of them were rather intuitive. For example, the variable CHECKUP1 (frequency of doctor visits) has a strong positive correlation with the cost problem—the less often a person visits a doctor the less is a chance he has cost problem. Employment status has a significant effect on access to health care. Retired people has an 83 % increase chance of facing cost problem in health care while employed has a 33 % decrease in probability to have cost problem. One of the most important variable is the presence of insurance. If a respondent reported that he has no insurance plan then the chance to have a limitation to health increases by 110 %.

A number of variables reported are not that intuitive, even counter-intuitive. If someone reported that he has income of $50,000 or more then this person got a probability of limitation in health care increased by almost 100 %. On the other hand, poor people whose income is lower than $25,000 has a 30 % decrease in probability of getting some cost problem. Another variable that showed unexpected result is gender. Males have a 24.3 % increase chance to have some issues with access to health care because of costs.

Table 6 Regression estimates

	Parameter		Estimate	Pr > ChiSq	Exp (Est)
	Intercept		−0.1908	0.0046	0.826
	CPDEMO1	1	0.0686	<0.0001	1.071
	CPDEMO1	2	0.0245	0.0433	1.025
	IMP_CHCCOPD	1	−0.0596	<0.0001	0.942
	IMP_CHCVISON	1	−0.1560	0.0019	0.856
	IMP_CHCVISON	2	0.0684	0.1671	1.071
	IMP_CHCVISON	7	−0.0731	0.2503	0.929
*	IMP_CHECKUP1	1	0.5354	<0.0001	1.708
	IMP_CHECKUP1	2	−0.0500	0.001	0.951
*	IMP_CHECKUP1	3	−0.2578	<0.0001	0.773
	IMP_CVDINFR4	1	−0.0559	0.0002	0.946
	IMP_EMPLOY	1	−0.1083	<0.0001	0.897
	IMP_EMPLOY	2	−0.1906	<0.0001	0.826
*	IMP_EMPLOY	3	−0.3963	<0.0001	0.673
*	IMP_EMPLOY	4	−0.3252	<0.0001	0.722
	IMP_EMPLOY	5	0.1249	<0.0001	1.133
	IMP_EMPLOY	6	0.1312	0.0005	1.140
*	IMP_EMPLOY	7	0.6039	<0.0001	1.829
	IMP_FLUSHOT5	0	0.1034	<0.0001	1.109
*	IMP_GENHLTH	1	0.4887	<0.0001	1.630
*	IMP_GENHLTH	2	0.2416	<0.0001	1.273
	IMP_GENHLTH	3	−0.0585	<0.0001	0.943
*	IMP_GENHLTH	4	−0.3139	<0.0001	0.731
*	IMP_HLTHPLN1	0	0.7430	<0.0001	2.102
	IMP_MENTHLTH		−0.0256	<0.0001	0.975
	......		......	......	......
*	SEX		0.2176	<0.0001	1.243
	......		......	......	......

Note Parameters marked with a * are the most significant; the number next to the parameter name indicates the categorical value of the parameter

6.2 Neural Networks

There is no simple method for selecting the best network from the enormous number of possible architectures. Multilayer perceptrons are black boxes with respect to interpretation. An additional analysis to understand the effects of the input variables to the target after the neural network model is built can be conducted to learn more insight about the neural network model. The chief benefit of neural networks for predictive modeling is their flexibility in representing universal approximation. The result of applying SAS EM neural network to the data is given in Table 7.

Table 7 Classification results for the neural network model

True status	Classification decision	
	Positive	Negative
True	128,883	5,425
False	14,115	2,899

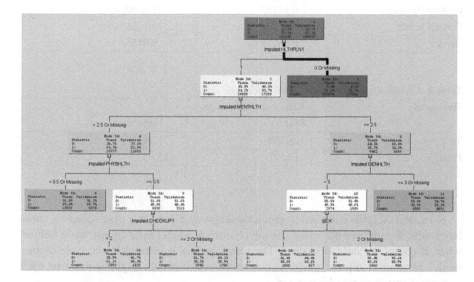

Fig. 3 Decision tree

6.3 Decision Tree

In general, the Decision Tree Model does not fit very well for this data set, partially due to the fact that the inputs are not highly associated with the target. However, according to Fig. 3 we can rely on number of rules that can give explanation about who is getting limitation to health care and who is not.

The *importance* of the variable, which SAS EM uses as a measures to decide the tree-splitting process (a variation of information gain) is given in Table 8.

6.4 K-nearest Neighbor

K-nearest neighbor classification is one of the main and simple classification methods and should be one of the first choices for a study when there is few information about the distribution of the data. The classification results on the data set is given in Table 9.

Table 8 Variable importance in the decision tree model

Variable	Importance measure	
	Training data	Validation data
Insurance presence	1	1
Mental health	0.36777	0.37241
Physical health	0.21028	0.21004
Routine checkup	0.1415	0.10948
General health	0.13724	0.13635
Gender	0.06219	0.04507

Table 9 Classification table for the MBR model

True status	Classification decision	
	Positive	Negative
True	128,750	1,343
False	18,197	3,032

K-nearest Neighbor is used to predict unknown values for a case based on similarity with K most similar cases. However, current model falls significantly behind when it comes to validation data. The training data is critical in the case of this model. In order to improve performance, transformation data might be required such as normalization or standardization.

7 Related Work

The early models for the use of health services were proposed in the 1960s [1] that identifies the factors affecting the use of health services. These models gradually evolved into an emerging model [2] as briefly described in Sect. 2.

Data mining has been widely applied to many fields, particularly in business. It has become increasingly popular only in recent years in healthcare, and was applied to various topics such as prediction of heart attack [15], classification on a medical data set of diabetic patients [10], and health social networks and consumer personalized medicine [16]. A recent comprehensive survey of data mining techniques and their applications in healthcare can be found in [19].

The survey data of the Behavioral Risk Factor Surveillance System has often been used as the standard data sets for health data analysis. Researchers have used the BRFSS data sets for tasks such as diet and physical activity behaviors [3], economic analysis of adult obesity [6], age-related disparities in cancer screening [9], etc.

8 Conclusion

In this paper, we applied different methodologies to predict access to health services on the survey data of Behavioral Risk Factor Surveillance System 2011. Four models were implemented to find which one has the most predictable results. Comparison results (ROC curves) show that Regression and Neural network models deliver the most explaining results while K-nearest Neighbor and Decision tree models fall behind based on misclassification rate.

Access to health services can be predicted by research models. Some variables show promise such as presence of insurance, physical and mental well-being, employment status, gender, etc. We plan to select another target variable to examine the performance of these models.

References

1. Andersen, R.M.: Behavioral Model of Families' Use of Health Services, vol. 25 (Center of Health Administration Studies, Chicago, 1968)
2. Andersen, R.M.: Revisiting the behavioral model and access to medical care: does it matter? J. Health Soc. Behav. 36(1), 1–10 (1995)
3. Bish, C.L., Blanck, H.M., Serdula, M.K., Marcus, M., Kohl, H.W., Khan, L.K.: Diet and physical activity behaviors among americans trying to lose weight: 2000 behavioral risk factor surveillance system. Obes. Res. 13(3), 596–607 (2005)
4. Centers for Disease Control and Prevention: Behavioral risk factor surveillance system survey data. http://www.cdc.gov/brfss/annual_data/annual_data.htm (2011)
5. Centers for Disease Control and Prevention: Physical activity and health. http://www.cdc.gov/physicalactivity/everyone/health/index.html (2013)
6. Chou, S.Y., Grossman, M., Saffer, H.: An economic analysis of adult obesity: results from the behavioral risk factor surveillance system. J. Health Econ. 23(3), 565–587 (2004)
7. Han, J., Kamber, M., Pei, J.: Data Mining Concepts and Techniques, 3rd edn. (Morgan Kaufmann, Waltham, 2012)
8. Hargraves, L.J., Hadley, J.: The contribution of insurance coverage and community resources to reducing racial/ethnic disparities in access to care. Health Serv. Res. 38(3), 809–829 (2003)
9. Jerant, A.F., Franks, P., Jackson, J.E., Doescher, M.P.: Age-related disparities in cancer screening: analysis of 2001 behavioral risk factor surveillance system data. Ann. Fam. Med. 2(5), 481–487 (2004)
10. Kaur, H., Wasan, S.K.: Empirical study on applications of data mining techniques in healthcare. J. Comput. Sci. 2(2), 194–200 (2006)
11. Koh, H.C., Tan, G.: Data mining applications in healthcare. J. Healthc. Inf. Manag. 19(2), 64–72 (2011)
12. Kuhn, M., Johnson, K.: Applied Predictive Modeling (Springer, New York, 2013)
13. Sarma, K.S.: Predictive Modeling With SAS Enterprise Miner: Practical Solutions for Business Applications (SAS, USA, 2007)
14. SAS: Getting Started with SAS Enterprise Miner 7.1. (SAS, USA, 2011)
15. Srinivas, K., Rani, B.K., Govrdha, A.: Applications of data mining techniques in healthcare and prediction of heart attacks. Int. J. Comput. Sci. Eng. 2(2) (2010)
16. Swan, M.: Emerging patient-driven health care models: an examination of health social networks, consumer personalized medicine and quantified self-tracking. Int. J. Environ. Res. Public Health 6(2), 492–525 (2009)

17. Wiltshire, J.C., Person, S.D., Kiefe, C.I., Allison, J.J.: Disentangling the influence of socioeconomic status on differences between african american and white women in unmet medical needs. Am. J. Public Health 9(99), 1659–1665 (2009)
18. Wu, J., Coggeshall, S.: Foundations of Predictive Analytics (CRC Press, Chicago, 2012)
19. Yoo, I., Alafaireet, P., Marinov, M., Gopidi, K.P.H.R., Chang, J.F., Hua, L.: Data mining in healthcare and biomedicine: a survey of the literature. J. Med. Syst. 36(4), 2431–2448 (2012)

Object Tracking Method Using PTAMM and Estimated Foreground Regions

So Hayakawa, Shinji Fukui, Yuji Iwahori, M.K. Bhuyan
and Robert J. Woodham

Abstract This chapter proposes a new approach for tracking moving objects in videos taken by a hand-held camera. The proposed method is based on the particle filter. The method is robust to occlusion by other objects. The 3D point map calculated by the Parallel Tracking and Multiple Mapping (PTAMM) is used for obtaining the positional relation between the target object and other moving objects. This causes improving the accuracy of the judgement of occlusion and being able to track the target object robustly when it is hidden by the others. The method uses the estimated foreground regions for calculating a part of likelihood. This increases the robustness of the tracking when the camera moving with rotation is used. The effectiveness of the proposed method is shown through the experiments using real videos.

Keywords Object tracking · Particle filter · PTAMM · 3D point map

S. Hayakawa · Y. Iwahori (✉)
Chubu University, Kasugai, Aichi 487-8501, Japan
e-mail: iwahori@cs.chubu.ac.jp

S. Hayakawa
e-mail: hayakawa@cvl.cs.chubu.ac.jp

S. Fukui
Aichi University of Education, Kariya, Aichi 448-8542, Japan
e-mail: sfukui@auecc.aichi-edu.ac.jp

M.K. Bhuyan
Indian Institute of Technology, Guwahati 781039, India
e-mail: mkb@iitg.ernet.in

R.J. Woodham
University of British Columbia, Vancouver, BC V6T 1Z4, Canada
e-mail: woodham@cs.ubc.ca

© Springer International Publishing Switzerland 2015   205
R. Lee (ed.), *Software Engineering Research, Management and Applications*,
Studies in Computational Intelligence 578, DOI 10.1007/978-3-319-11265-7_16

1 Introduction

The object tracking method is a significant technique in the field of computer vision. The method can be applied to the traffic surveillance system, the behavior recognition, and so on. Many methods based on particle filter (PF) have been proposed recently to track moving objects robustly [3, 6–12]. The particle filter based tracking is robust to occlusion and noise.

Appearance information, such as color and edge information of the object, is used to calculate the likelihood. Methods using appearance information tend to fail in tracking the target object when objects similar to the target object come close to the target, when they intersect with it or when they hide it.

Methods which are robust to intersection with similar objects have been proposed [1, 12]. The method [1] is proposed for human detection and tracking. It detects human body parts by the boosting method learned in advance, finds humans by combining the detected parts and tracks them by using the information of the body parts. The method can track only human. The method [12] introduced likelihood calculated by the probabilistic background model so that the method may become a robust tracking to the intersection with similar objects. Increasing the accuracy of tracking, and reducing the processing cost remain as the future work.

In this chapter, a new approach for tracking objects in a video sequence taken by a moving camera is proposed. The proposed method obtains the three dimensional positions of the objects. The position in the 3D space is obtained by the 3D point map calculated by Parallel Tracking and Multiple Mapping (PTAMM) [2]. The positional relation of the target object and other moving objects can be obtained by the positions on the map. This increases the accuracy of judging whether the target object is hidden by the others or not. Furthermore, the likelihood function is improved. The likelihood which is calculated by feature points on moving objects or around it is used. This can reduce the likelihood of particles which cannot catch the target object. The likelihood function increases the robustness of tracking when the camera which moves with rotation is used. Results are demonstrated by experiments using real video sequences.

2 Parallel Tracking and Multiple Mapping

The proposed method uses the 3D point map calculated by the PTAMM to obtain the positional relation between the moving objects. The PTAMM is a method for the Augmented Reality. The method uses no marker and can estimate the position and the direction of a hand-held camera in real-time by using feature points in the images.

Fig. 1 Example of 3D point map

The outline of the procedure of the PTAMM is shown below.

1. Initialize
2. Tracking
3. Updating map

(2) and (3) are iterated and are processed in parallel.

At (1), the 3D point map is initialized by the five-point stereo algorithm [10]. At the tracking step, the camera pose relative to the 3D point map is estimated. At (3), the feature points in the frame is added to the 3D point map when the frame satisfies the conditions as the keyframe. Otherwise, the 3D point map is improved by the bundle adjustment. The example of the 3D point map generated by the PTAMM is shown in Fig. 1.

3 Proposed Object Tracking Method

The proposed method is a tracking method based on PF. PF is a method for estimating internal state variables by observed data and the previous state variables. PF approximates the posterior probability density for the state variables by many particles. Each particle is a weighted sample of the state variables. The posterior probability density is represented by the weights of the particles.

PF estimates the state variables by the following steps. First, the state variables at time t are predicted by the variables at $t - 1$ and a state transition model. Next, the weight of each particle is obtained by the likelihood calculated by the likelihood function. As a result, the discrete approximation of the posterior density is obtained. At last, particles are weighted resampling with replacement. These three steps are iterated.

The state variables, the state transition model and the likelihood function for the proposed method are described in the following.

3.1 State Variables

Let the state variables of i-th particle at t be represented by $c_t^{(i)}$. $c_t^{(i)}$ is defined as follows:

$$c_t^{(i)} = [x_t^{(i)}, y_t^{(i)}, u_t^{(i)}, v_t^{(i)}, w_t^{(i)}, h_t^{(i)}]^\top \tag{1}$$

where $(x_t^{(i)}, y_t^{(i)})$ represents the center position of the tracking object on the image, $(u_t^{(i)}, v_t^{(i)})$ represents the velocity of the tracking object, and $w_t^{(i)}$ and $h_t^{(i)}$ represent the width and height of the rectangle which involves the tracking object.

When the target object is hidden by other moving objects, the velocity and the size of the target object cannot be obtained. In such a case, $u_t^{(i)}, v_t^{(i)}, w_t^{(i)}$ and $h_t^{(i)}$ are not used. How to judge whether the target object is hidden or not is described in the next section.

3.2 State Transition Model

The proposed method defines two transition model. One is for the case that the target object is hidden by other moving objects and the other is for other cases.

The state transition model for the former case is defined by the following equation.

$$c_t^{(i)} = \begin{cases} c_{t-1}^{(i)} + g_t^{(i)} & (\text{if } L_{t-1}^{(i)} > \overline{L}_{t-1}) \\ (\tilde{x}^{(i)}, \tilde{y}^{(i)}, \mathbf{0}^\top)^\top + g_t^{(i)} & (\text{otherwise}) \end{cases} \tag{2}$$

where $g_t^{(i)}$ means a Gaussian noise vector, $L_{t-1}^{(i)}$ means the likelihood value of i-th particle, $\overline{L}_{t-1}$ represents the mean of the likelihood. $\tilde{x}^{(i)}$ is selected randomly in $x_{t-1}^{(i)}, x_{o,t}' - \hat{w}_{o,t-1}/2$ and $x_{o,t}' + \hat{w}_{o,t-1}/2$. $\tilde{y}^{(i)}$ is selected randomly in $y_{t-1}^{(i)}$ and $\hat{h}_{o,t-1}/2$. $(x_{o,t}', y_{o,t}')$ means the temporary position of the object which hides the target object and they are calculated by the following equation.

$$x_{o,t}' = \hat{x}_{o,t-1} + \hat{u}_{o,t-1}, \quad y_{o,t}' = \hat{y}_{o,t-1} + \hat{v}_{o,t-1} \tag{3}$$

where $(\hat{x}_{o,t-1}, \hat{y}_{o,t-1})$ means the position of the object estimated at $t-1$ and $(\hat{u}_{o,t-1}, \hat{v}_{o,t-1})$ means the estimated velocity. $\hat{w}_{o,t-1}$ and $\hat{h}_{o,t-1}$ mean the estimated width and height of the rectangle. The function makes $(x_t^{(i)}, y_t^{(i)})$ exist inside the rectangle of the other object or around it. The method can track the target object again when it reappears from somewhere around the object occluding the target object.

The other transition model is defined by Eq. (4).

$$c_t^{(i)} = c_{t-1}^{(i)} + (u_{t-1}^{(i)}, v_{t-1}^{(i)}, \mathbf{0}^\top)^\top + g_t^{(i)} \tag{4}$$

3.3 Likelihood Function

The proposed likelihood function is defined by Eq. (5):

$$L_t^{(i)} = L_{t,c}^{(i)} \, L_{t,v}^{(i)} \, L_{t,d}^{(i)} \, L_{t,f}^{(i)} \tag{5}$$

where $L_{t,c}^{(i)}$, $L_{t,v}^{(i)}$, $L_{t,d}^{(i)}$ and $L_{t,f}^{(i)}$ mean the likelihood based on the color, that based on the velocity, that based on the distance between the objects and that based on the estimated foreground regions respectively.

Each likelihood is described in the following.

3.3.1 Likelihood Based on Color

$L_{t,c}^{(i)}$ is calculated by the following equation.

$$L_{t,c}^{(i)} = \exp\left(k_c\left(\frac{1}{n-2}\sum_{j=2}^{n-1}S_j^{(i)}\right)^2\right) \tag{6}$$

where k_c means the proportionality factor and $S_j^{(i)}$ means the color similarity. $S_j^{(i)}$ is calculated by the following process. First, the reference image of the target object, which is obtained in advance, and the rectangle region, of which the center position, width and height are $(x^{(i)}, y^{(i)})$, $w^{(i)}$ and $h^{(i)}$, are divided into n blocks horizontally. Next, $S_j^{(i)}$ is calculated by Swain's histogram intersection [11] using the histogram of j-th region and its neighbors of the reference image and that of the corresponding regions of j-th rectangle. The joint histograms of HS histogram and SV histogram are used at this time. The number of bins of the HS and SV histogram is set to 10.

The proposed method changes the value of k_c according to the situation. When the target object is close to the similar objects, it is better that L_c is small. Otherwise, the proposed tracker may track the other object because the likelihood of particles of which $(x_t^{(i)}, y_t^{(i)})$ exists in the region of the other object becomes high. k_c is changed according to the similarity of color between the target object and other objects.

k_c is given by Eq. (7).

$$k_c = k \, \underset{l \in \{1, \cdots, M\}}{\arg\min} \, ((1.0 - S_c^l)) \tag{7}$$

where k means a constant, M means the number of the objects which exist near the target object and S_c^l means similarity of color between the reference color histogram of the target object and l-th object. k is set to be 20 according to [9].

3.3.2 Likelihood Based on Velocity

$L_{t,v}^{(i)}$ is calculated by the following equation.

$$L_{t,v}^{(i)} = \exp\left(-\frac{\sqrt{(x_t^{(i)} - x_t')^2 + (y_t^{(i)} - y_t')^2}}{2\sigma_v^2}\right) \tag{8}$$

$$\sigma_v = \sqrt{\hat{u}_{t-1}^2 + \hat{v}_{t-1}^2 + 1}$$

where (x_t', y_t') means the temporary position of the target object and $(\hat{u}_{t-1}, \hat{v}_{t-1})$ means the velocity estimated by the particle filter at $t-1$. x_t' and y_t' are obtained by the same manner as Eq. (3). $L_{t,v}^{(i)}$ becomes larger when $(x_t^{(i)}, y_t^{(i)})$ is closer to (x_t', y_t').

σ_v tunes the variance of the exponential. It is defined so that it may be changed according to the velocity of the target object. When the target object moves faster, the area in which the target object may exist is wider. σ_v becomes larger so that likelihood of more particles becomes large in this case. In contrast, σ_v becomes small when the target object moves slowly.

3.3.3 Likelihood Based on Distance

$L_{t,d}^{(i)}$ is calculated by the following equation.

$$L_{t,d}^{(i)} = \begin{cases} 1 & \text{(if no object exists near the target)} \\ \prod_{l=1}^{M}\left(\varsigma_{\alpha_{t,l}^{(i)}}(D_{t,l}^{(i)} - D_{t,l})S_{t,l} + (1.0 - S_{t,l})\right) & \text{(otherwise)} \end{cases} \tag{9}$$

where M means the number of the objects existing around the target object, $\varsigma(\cdot)$ means the sigmoid function, $\alpha_{t,l}^{(i)}$ means the gain of $\varsigma(\cdot)$, $D_{t,l}^{(i)}$ means the distance between $(x_t^{(i)}, y_t^{(i)})$ and the temporary position of l-th object and $D_{t,l}$ means the distance between the temporary positions of the target object and l-th object. $S_{t,l}$ is the cube root of the similarity calculated by the Histogram Intersection between the target object and l-th object. Figure 2 shows $\alpha_{t,l}^{(i)}$, $D_{t,l}$ and $S_{t,l}$ and those effects on $L_{t,d}^{(i)}$.

$L_{t,d}^{(i)}$ is introduced to prevent the method from tracking other objects mistakenly when the target object intersects with them. It becomes smaller as $(x_t^{(i)}, y_t^{(i)})$ is closer to the other object. On the other hand, $L_{t,d}^{(i)}$ becomes larger as $S_{t,l}$ becomes smaller. This prevents $L^{(i)}$ from becoming too small and causes the stable tracking when $S_{t,l}$ is small.

Fig. 2 Parameters of $L_{t,d}^{(i)}$ and those effect on $L_{t,d}^{(i)}$

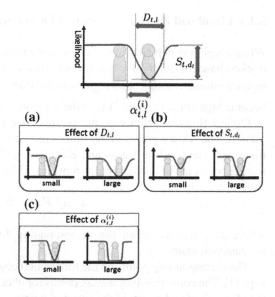

$\alpha_{t,l}^{(i)}$ is determined by the following equation.

$$\alpha_{t,l}^{(i)} = \frac{\min(w_t^{(i)}, h_t^{(i)}) - \min(w_{t-1}^{(i)}, h_{t-1}^{(i)})}{6\sigma_{w|h}} + k_f \tag{10}$$

where $\sigma_{w|h}$ means the variance of the Gaussian noise for the smaller variable in $w_{t-1}^{(i)}$ and $h_{t-1}^{(i)}$. The noise is used for the state transition. k_f is a constant for determining the minimum value of $\alpha_{t,l}^{(i)}$. It is set to 0.05 in the experiments. $\alpha_{t,l}^{(i)}$ is introduced to prevent the rectangle size estimated by the proposed method from becoming too large. $L_{t,c}^{(i)}$ of the particle which has large rectangle size tends to become large when the target object intersects with others. $\alpha_{t,l}^{(i)}$ makes $L_t^{(i)}$ of the particle of which the rectangle size is large small.

The method should know how many objects exist around the target object. It is assumed that all moving objects in the image are tracked by the particle filter. The method recognizes the existence of the l-th object around the target object when the temporary position of the l-th object exists in the circle of which the center position is (x_t', y_t') and the radius is $6\sqrt{\sigma_x^2 + \sigma_y^2}$, where σ_x and σ_y mean the standard deviations which are used for generating the Gaussian noises for adding to $(x_t^{(i)}, y_t^{(i)})$.

3.3.4 Likelihood Based on Estimated Foreground Object Regions

When a region which has similar color to the target object exists in the background, trackers based on color information tend to track the region mistakenly. The proposed method estimates foreground regions in the frame. $L_{t,f}^{(i)}$ is defined so that $L_{t,f}^{(i)}$ may become high when $(x_t^{(i)}, y_t^{(i)})$ is in the estimated foreground region.

Optical flows of feature points are used for the estimation of the foreground regions. The feature points in the frames at t and $t-1$ are obtained and the optical flows are calculated by Lucas and Kanade [8]. The relation of the corresponding feature points in the background is represented by Eq. (11).

$$x_{t-1}^{\top} F x_t = 0 \tag{11}$$

where x_{t-1} and x_t mean the corresponding feature points and F means the fundamental matrix.

The corresponding points on the foreground objects or around them do not satisfy Eq. (11). The corresponding feature points for obtaining F are selected by RANSAC [5]. F is calculated by using the selected points.

After F is obtained, the values of left side of Eq. (11) for all pairs of the corresponding feature points are calculated. The points of which the values are close to 0 are regarded as the points in the background. The feature points which have larger values than the threshold are regarded as the points on the foreground object and they are selected. The example of the result of this process is shown in Fig. 3.

The probability $p(x_t)$ that x_t belongs to the foreground region is calculated by the result of the above process. $L_{t,f}^{(i)}$ is calculated by $p(x_t)$. How to calculate $p(x_t)$ and $L_{t,f}^{(i)}$ is described in the following.

The circles of which the center points are the selected feature points are considered. At each pixel, the number of the circles which include the pixel is counted.

Fig. 3 Result of selecting process. The *gray* points are the feature points which are regarded as the points in the background and the *white* points are the selected feature points

The probability that each pixel belongs to foreground region is calculated by normalizing the value obtained through the above process.

The radius r of the circle is calculated by the Eq. (12).

$$r = k_r + k_r * p(x_t) \tag{12}$$

where k_r means a constant.

After obtaining the probability, the image segmentation method [4] is applied to the t-th frame. $L_{t,f}^{(i)}$ is calculated by Eq. (13).

$$L_{t,f}^{(i)} = \begin{cases} \frac{1}{N_{\mathbf{R}^{(i)}}} \sum_{x \in \mathbf{R}^{(i)}} p(x) & \text{(if } M > 0) \\ 1 & \text{(if } M = 0) \end{cases} \tag{13}$$

where $\mathbf{R}^{(i)}$ means the segmented region where $(x_t^{(i)}, y_t^{(i)})$ exists, and $N_{\mathbf{R}^{(i)}}$ means the number of pixels in $\mathbf{R}^{(i)}$.

4 Judgement of Occlusion

The proposed method judges whether the target object is hidden by the other objects or not. It is recognized that the target object is occluded by the other object when the following two conditions are satisfied at the same time.

- Not less than 80 % region of the rectangle for the target object overlaps the rectangle for the other objects
- The target object exists farer from the camera than the other moving object

The positional relation between the target object and the other objects is obtained by the 3D point map calculated by the PTAMM. It is assumed that XY plane of the 3D point map is the same plane as the ground and that the bottom of the rectangle which consists of $(x_t^{(i)}, y_t^{(i)})$, $w_t^{(i)}$ and $h_t^{(i)}$ is on the ground. Let the center point of the bottom of the estimated rectangle for the target object at t be represented by $x_{t,c}$ and let that for the l-th object be represented by $x_{t,c}^l$. How to obtain the positional relation is described in the following.

First, $x_{t-1,c}$ and $x_{t-1,c}^l$ are mapped to the 3D point map. Next, the temporary position of the target object $x_{t,c}'$ and that of the l-th object $x_{t,c}^{l'}$ on the map is calculated by the following equations.

$$x_{t,c}' = 2x_{t-1,c} - x_{t-2,c} \tag{14}$$

$$x_{t,c}^{l'} = 2x_{t-1,c}^l - x_{t-2,c}^l \tag{15}$$

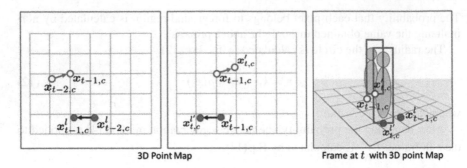

Fig. 4 Estimation of temporary position on 3D point map. The vectors of the same color have the same distance and the same length. The sizes estimated by the particle filter at $t-1$ are used as the sizes of the *rectangles*

The distance between $x'_{t,c}$ and the camera and that between $x^{l'}_{t,c}$ and the camera are calculated and the positional relation of the target object and the l-th object is determined. The position of the camera can be obtained by the PTAMM.

Figure 4 shows the judgement process.

5 Experiments

The experiments using real videos were done to confirm the effectiveness of the proposed method. Four videos (Scene1, Scene2, Scene3 and Scene4) were used in the experiments. All scenes were taken by a hand-held camera. In the Scene1 and the Scene4, three persons with similar clothes intersected with each other. The Scene2 and the Scene3 are taken by a camera which moves with rotation. The image size of each frame was 720×480.

The number of the particles for tracking an object was set to 100. While, it was set to 300 when the similar objects existed near the target object. k_r in Eq. (12) is set to 20. The regions of the objects were given manually and the method was initialized by the regions because the objects existed from the first frame.

Figure 5 shows the examples of the frames used in the experiments. The tracking results of the proposed method are shown in Fig. 6. The rectangles in the images mean the estimated position of the target objects. The white rectangle means that the target object is hidden by the other objects. The figures show that the proposed method can track robustly at all scenes.

The experiments by the method [12] were done for comparison with the proposed method. The results are shown in Fig. 7.

Scene1

Scene2

Scene3

Scene4

Fig. 5 Examples of input images

The particle filter uses the Gaussian noises at the prediction step. The method based on the particle filter can fail in tracking when the method is applied many times to the same video. The success rates of tracking were calculated for the evaluation of the robustness. The proposed method and the method [12] were applied 25 times to each scene. It is judged that the tracking was success when the method continued to track the target object from the start to the end. The rates are shown in Table 1.

The proposed method can get better rates at all scenes than the method [12]. Especially, the results of the proposed method for Scene1 and Scene3 are widely improved. Two persons are occluded by one person in Scene1. The method [12] cannot distinguish the persons. The proposed method obtains the positional relation

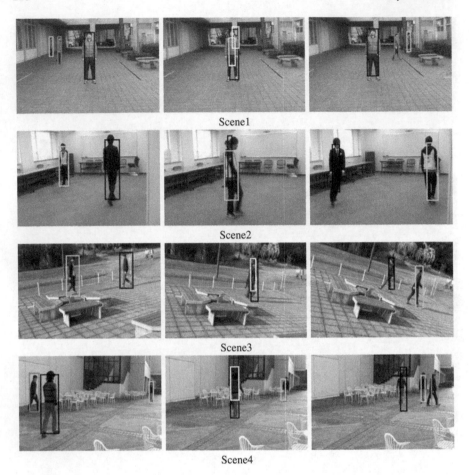

Scene1

Scene2

Scene3

Scene4

Fig. 6 Results of proposed method

of the persons in the 3D space and can distinguish them. The movement of the camera by which Scene3 is taken includes the rotation. The method [12] cannot make the background model well in such a scene. The proposed method can estimate the foreground regions by the epipolar constraint.

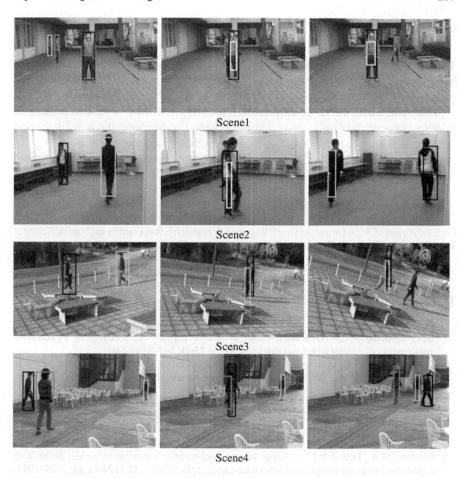

Scene1

Scene2

Scene3

Scene4

Fig. 7 Results of method [12]

Table 1 Success rates of tracking

	Scene1	Scene2	Scene3	Scene4
Proposed method (%)	84	68	64	92
Method [12] (%)	40	40	13	88

6 Conclusion

This chapter proposed the new approach for tracking objects in videos taken by a hand-held camera. The proposed method is robust to the scene which is taken by the camera moving with rotation and to occlusion by the other moving objects.

The method uses the 3D point map calculated by PTAMM to obtain the positional relation between the target object and the others. Furthermore, the new likelihood function is introduced. These causes improving the accuracy of the tracking.

More effective use of the 3D point map and improvement of accuracy of tracking remain as the future work.

Acknowledgments Fukui's research is supported by JSPS Grant-in-Aid for Young Scientists (B) (23700199). Iwahori's research is supported by JSPS Grant-in-Aid for Scientific Research (C) (26330210) and a Chubu University Grant. Woodham's research is supported by the Natural Sciences and Engineering Research Council (NSERC).

References

1. Bo, W., Nevatia, R.: Detection and tracking of multiple, partially occluded humans by Bayesian combination of edgelet based part detectors. IJCV **75**(2), 247–266 (2007)
2. Castle, R., Klein, G., Murray, D.W.: Video-rate localization in multiple maps for wearable augmented reality. In: 12th IEEE International Symposium on Wearable, Computing, pp. 15–22 (2008)
3. Del Bimbo, A., Dini, F.: Particle filter-based visual tracking with a first order dynamic model and uncertainty adaptation. Comput. Vis. Image Underst. **115**(6), 771–786 (2011)
4. Felzenszwalb, P.F., Huttenlocher, D.P.: Efficient graph-based image segmentation. IJCV **59**(2), 167–181 (2004)
5. Fischler, M.A., Bolles, R.C.: Random sample consensus: a paradigm for model fitting with applications to image analysis and automated cartography. Comm. ACM **24**(6), 381–395 (1981)
6. Isard, M., Blake, A.: CONDENSATION—conditional density propagation for visual tracking. IJCV **29**(1), 5–28 (1998)
7. Khan, Z., Balch, T., Dellaert, F.: An MCMC-based particle filter for tracking multiple interacting targtes. In: Proceeding of ECCV2004, vol. 4, pp. 279–290 (2004)
8. Lucas, B., Kanade, T.: An iterative image registration technique with an application to stereo vision. In: Proceedings of IJCAI, pp. 674–679 (1981)
9. Perez, P., Hue, C., Vermaak, J., Gangnet, M.: Color-based probabilistic tracking. In: Proceedinf of 7th ECCV, Vol. 1, pp. 661–675 (2002)
10. Stewenius, H., Engels, C., Nister, D.: Recent developments on direct relative orientation. ISPRS J. Photogrammetry Remote Sens. **60**(4), 284–294 (2006)
11. Swain, M.J., Ballard, D.H.: Color indexing. IJCV **7**, 11–32 (1991)
12. Watanabe, G., Fukui, S., Iwahori, Y., Bhuyan, M.K., Woodham, R.J.: Robust tracking method based on particle filter for crossing of targets with similar appearances. In: SCIS-ISIS 2012, F1–54-4, pp. 1–4 (2012)

Application of Detecting Blinks for Communication Support Tool

Ippei Torii, Kaoruko Ohtani, Shunki Takami and Naohiro Ishii

Abstract In this paper, we describe a new application oper-ated with blinks for physically handicapped children who cannot speak to communicate with others. Process of detecting blinks is performed in the following steps. (1) To detect an eye area (2) To distinguish opening and closing of eyes (3) To add the method using saturation to detect blinks (4) To decide by a conscious blink (5) To improve the accuracy of detection of blinks We reduce the error to detect blinks and pursue the high precision of the eye chasing program. The degree of disablement is varied in children. So we develop the system to be able to be customizes depends on the situation of users. And also, we develop the method into a communication application that has the accurate and the high-precision blink determination system to detect letters and put them into sound.

Keywords Voice output communication aid (VOCA) · Physically handicapped children · OpenCv · Haar-like eye ditection

1 Introduction

Special support schools in Japan need some communication assistant tools especially for physically handicapped children. In this study, physically handicapped children are defined as children with permanent disablements of their trunks and limbs because of cerebral palsy, muscular dystrophy, spinal bifida and so on. Their body movements are very limited and many of them also have mental disorders, so they

I. Torii (✉) · K. Ohtani · S. Takami · N. Ishii
Department of Information Science, Aichi Institute of Technology, Aichi, Japan
e-mail: mac@aitech.ac.jp

K. Ohtani
e-mail: ruko2011@gmail.com

S. Takami
e-mail: mail@takamin.net

N. Ishii
e-mail: ishii@aitech.ac.jp

© Springer International Publishing Switzerland 2015
R. Lee (ed.), *Software Engineering Research, Management and Applications*,
Studies in Computational Intelligence 578, DOI 10.1007/978-3-319-11265-7_17

cannot communicate with their families or caregivers. It prevents helpers from understanding what they really need or think.

Taking the situation into consideration, we develop a communication support tool operated by blinks for physically handicapped children. We use front cameras on tablets (iPad, iPad mini of Apple Inc. iOS5.0). After several verifications and examination, we have developed a contactless communication assistant application "Eye Talk". The application does not malfunction by surroundings, such as brightness or differences of eye shapes.

2 Purpose of the Study

The most common way to communicate with physically handicapped children is using "Yes=○" or "No=×" cards. (Figure 1) For example, if a caregiver wants to ask a child whether he/she wants to drink water, the caregiver will ask him/her "Do you want to drink water?" and show him/her cards with ○ and × by turns. If the child takes a look at ○ card or put his/her eye on it longer than the other one, the caregiver will know he/she may want to drink water.

But caregivers have to predict what patients need or want to say by their experiences or circumstances in this method. So the questions made by caregivers can be totally different from what patients really want to say. And also, it is difficult to figure out the movement of eyes of patients. Sometimes caregivers have to just guess the answer.

Some communication support tools using movements of eyes for these physically handicapped people have been released already, such as TalkEye [1] or Let's Chat [2]. Most of them are relatively expensive because they require some special equipment. For example, TalkEye requires the executive head set to measure the movement of eyes.

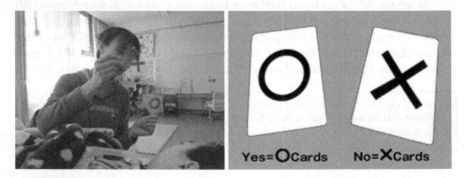

Fig. 1 Using "Yes=○" or "No=×" cards to communicate with a physically handicapped child

3 Structure of the System

Process of detecting blinks is performed in the following steps,

① To detect an eye area(By using Opencv Haar-like eye-detection)

② To distinguish opening and closing of eyes(By using the complexity of binarized image)

③ To add the method using saturation to detect blinks(Aiming more accurate detection)

④ To decide by a conscious blink(To define what is the "conscious blink")

⑤ To improve the accuracy of detection of blinks.

3.1 Detection of an Eye Area

There are many methods to detect an object. We choose OpenCv that is a library of programming functions for real time computer vision for image processing in this study. OpenCV's face detector uses a method that Viola [3] developed first and Lienhart [4] improved.

We use differences of brightness to figure out a face area and to remove other areas. Next, we detect an eye-area. Since an eye-area is inside a face area, we use data of a face area obtained by the method to obtain the average brightness of upper and lower part of eyes to define an eye-area.

Next, we detect an eye-area. Since an eye-area is inside a face area, we use data of a face area obtained by the method to obtain the average brightness of upper and lower part of eyes to define an eye-area.

3.2 Detection of Eye Opening and Closing

There are two methods to detect blinks. First one is calculating the size of the black area of eyes by using spiral labeling [5] and considering it as a blink when the size becomes smaller than the threshold. The second one is using the difference of the value of brightness of images to determine blinks. Figure 2 shows the image of spiral labeling. In spiral labeling, we search pixels from the starting pixel (1 in a square in Fig. 5), calculate the medial level of the difference, and count the pixels within the threshold. In this method, we can reduce the time of processing if we can get the starting pixel because the area of processing is limited.

But many physically handicapped children tend not to be able to open their eyes wide enough and their iris of the eye is relatively small, so it is difficult to detect the center of iris of the eye. So we use Value in HSV (Hue, Saturation, Value) in color space [6] to determine blinks. In this method, we cut an eye area based on the coordinate data obtained by Haar-like classifier and size it to reduce the load to a devise.

Fig. 2 Image of spiral labeling

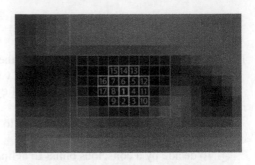

Fig. 3 Image of eye opening and closing

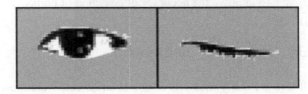

Then we obtain the average of brightness of the eye area. When someone closes his/her eyes, the average rises. We determine it as eye closing. Figure 3 shows the image of eye opening and closing.

Usually, the classrooms of the most of special support schools are relatively dark to avoid giving extra impetuses to children, so we cannot get enough amount of light. So it is difficult to determine blinks, because there is no difference of darkness around the eye area and the average of brightness of white and black part of eyes that is close to the brightness of skin.

3.3 Detecting Method by Saturation in Color Space

Here we develop the method using saturation in color space. In this method, we calculate the average of saturation (0–255 in saturation of HSV) of area C (the center of the iris) and W (white part of the eye) in Fig. 4. If the measured value is lower than the average of saturation, we determine it as eye closing. We collect many numbers of eye area data to calculate the average of saturation for eye opening.

Fig. 4 To obtain the average of saturation

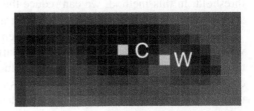

When a user starts the system, it calibrates the picture of eye opening. Based on this picture, the system determines it as eye closing when the saturation of eye area is lower than the threshold. We use the threshold (the average of saturation) between 5 flames to 14 flames before of the present flame. But on the other hand, it detects a small movement such as an unconscious blink or moving of the face. So we add complexity of image of the eye in last 3 flames including the present flame to exclude the error to determine blinks.

3.4 Detecting Method by Complexity of Image

We use the difference of the outlines between eye opening and closing. We determine it as eye closing when the pixel of the difference of amount of edge in the picture becomes flat. The threshold is based on the average value of the amount of edges of 10 flames (5 to 14 flames before of the present flame). We determine blinks with the difference between the threshold and the present flame. Figure 5 shows the utilization to each frame of the process complexity and saturation of the blink detection.

The purpose of adding this method is to stop to detect an unconscious blink or a movement of the face. Figure 6 shows the image of complexity of the image of the eye.

First, we try to find the most optimum value of combina-tion of the saturation and complexity to determine eye closing. The setting that has less error is $LP = 0.75$, $DP = 0.86$. We set this number as the threshold.

Figure 7 shows the image when the application starts. ABC shows the correlation values of complexity and DEF shows the correlation value of saturation. A is the value of complexity of the present flame. B is the average value of complexity of 5–14 flames before of the present flame. C is the value of complexity multiplying

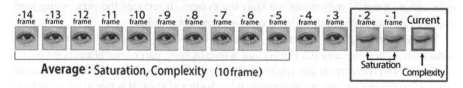

Average : Saturation, Complexity (10 frame)

Fig. 5 Flames for calibration

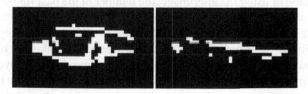

Fig. 6 Complexity of image of the eye

Fig. 7 Image and value of
eye area

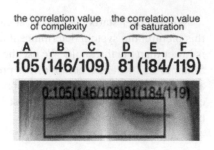

the threshold LP (0.75) to the value of the present flame. D is the value of saturation of the present flame. E is the average value of saturation of 5–14 flames before of the present flame. F is the value of saturation multiplying the threshold DP (0.86) to the value of the present flame. When B is larger than A, and E is larger than D, we deter-mine it as eye closing.

3.5 Developing Afterimage Method

After we developed the basic model of the system, we carried on a clinical experiment in the school for handicapped children. A subject suffering from spinal muscular atrophy has very weak muscle strength and cannot blink longer enough. So we need to improve the system to determine blinks even situations of users are different.

For users who can blink strongly enough, the method to determine blinks by the complexity and saturation is appropriate, but another detection method is required for users who do not have enough muscle strength. So we need to increase the sensitivity to detect blinks. Therefore, it is necessary to increase the processing power for detection of blinks. Since the tablet is fixed to the bed, we detect the position of eyes using OpenCV only when we start the system. It reduces the processing load and we can use 4–25 frames per second to determine blinks.

We use afterimages to determine weak and fast blinks instead of complexity and saturation. When the area (C, Fig. 8) that is overlapping part of the afterimage of iris (A, Fig. 8) and the current iris situation (B, Fig. 8) is diminishing rapidly comparing from a few frames before, we determine it as closing of eyes. It is possible to exclude a slow change and capture a rapid change.

By the method of using the afterimage, the number of the past frames to compare increases and we can capture changes with high accuracy. As a result, we can quantify the changes of series of movements "eye-opening → eye-closing → eye-opening". And it is possible to determine the situation of eyes as short blink, long blink, closing eyes continuously, except the malfunction due to fine movement of eyes. Figure 12 shows determination by afterimages visually. A is the afterimage to be compared, B is the iris in current frame, and C is the overlapped area of the afterimage and current frame.

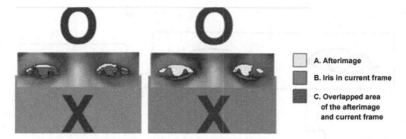

	A. Afterimage
	B. Iris in current frame
	C. Overlapped area of the afterimage and current frame

Fig. 8 Image and value of eye area

In order to handle with face color and brightness of light, calibration is automatically activated when the black part in the image is doubled. Calibration means to detect the position of eyes and to set a threshold for binarizing. If the value of binarization threshold is too high or too low, it cannot detect blinks. We perform calibration to obtain the correct value for the blink determination and set the threshold. As a result, it is possible to automatically capture the position of eyes in both a dark place and a bright place or even if a user moves his/her face.

The upper graph in Fig. 9 is a change of the difference of values between the number of pixels of the current frame and afterimage. The middle line is the threshold and we set to determine it as a blink when more than 1/18 of the total number of pixels is changed. When the value drops below the threshold after the value exceeds the threshold then certain number of frames (setting to five frames) are passed, it is determined as a blink. The vertical straight lines indicate the blink determination. It is possible to set the parameters not to react with too long or too fast blinks.

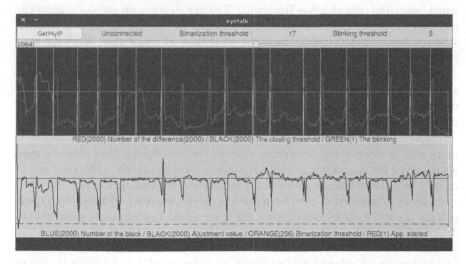

Fig. 9 The graph of change of the difference of values between the number of pixels of the current flame and afterimage

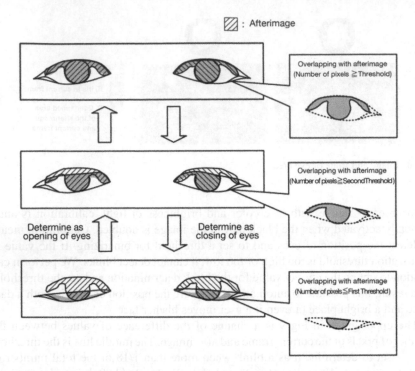

Fig. 10 Diagram of opening and closing of eye

The bottom part in Fig. 9 is a graph for the calibration. This shows the change of fractions of the pixel volume in the current frame. The dotted line is the threshold when binarizing (Binarization Threshold). The numbers are up and down automatically by the brightness of the room and the color of the skin to create an optimal state for detection. The middle line is the ideal value of calibration. When the state of opening eyes is close to the line, it is working properly.

It is possible to detect blinks using the value of the current frame, but if a user has a long eyelashes, the outline becomes complicated because it acquires the contour of the eyelashes, the value is unstable when the eyes are closing. Therefore, we designed the method of using the afterimage to correspond to blinks of any user.

Figure 10 shows how to determine the eye opening and closing. It is difficult for users who have weak muscle strength to close the eyes tightly. They close the eyes weakly even for the conscious blink. Therefore, their conscious blinks are tends to be only half-closed ones as shown in the second and third pictures of Fig. 10.

In the determination of eye closing in Step 4(S4) and Step 5(S5) on Fig. 11, the process is carried out by determining the value of the first threshold to be able to determine the eye closing even blinks are not strong enough. Sensitivity of the blink determination can be selected from 0 to 100 on the setting screen of the application. The higher the set value of the first threshold is larger (closer to 100), it is easy to determine eye closing of a user.

Fig. 11 Flowchart of eye
opening and closing

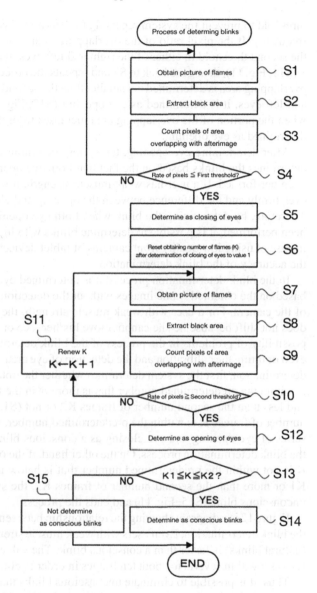

After determining eye closing, the obtained numbers of frames (K) is reset to 1 (S6, Fig. 11), then determine eye opening by capturing the process of the opening of eyes of a user. To determine eye opening, first, the system acquires a frame image captured by the camera (S7, Fig. 11), then extracts the black area indicating the pupil of right and left eyes (S8, Fig. 11). And it counts the number of pixels of the overlapping area with afterimage (S9, Fig. 11), then determines the ratio of the pixels of overlapping are in afterimage is equal or smaller than the second threshold (S10, Fig. 11). The value of the second threshold is determined slightly larger than the first threshold to prevent the system repeating to determine eye opening and closing too frequently. If the pixel ratio of the overlapping area is determined not smaller than the second threshold in either of the right and left eyes, the system increments K by 1 (S11, Fig. 11), then goes back to S7 and repeats the process. If the pixel ratio of the overlapping area is determined not smaller than the second threshold in both the right and left eyes, it is determined as eye opening (S12, Fig. 11). As shown in Fig. 10, when the portion of the overlapping area increases from the eye closing status, it is determined as eye opening.

After determining eye opening, the system determines blinks. Thus, the system determines the blinks based on the fact that both eye opening and eye closing have been per-formed. If a user has weak muscle strength, it is difficult to open or close eyes firmly and the difference between the opening and closing is not clear enough. Therefore, by determining the blink when both eye opening and eye clos-ing have been performed, it is possible to determine blinks with high accuracy. Further, even in case of using low quality front cameras of tablet devices, it is possible to improve the accuracy of the blink determination.

In the blink determination process, it is determined eye opening and eye closing based on the obtained frame images without the detection of eye area or calibration of the camera. For a user with weak muscle strength, the detected position of eyes does not shift because he/she cannot move his/her face or body a lot. Therefore, the possibility of problems in the process of the blink determination is relatively small even omitting the calibration and the detection of eye area. In the process of the blink determination, first the system determines whether the obtained number of frames K is within the predetermined number that is more than the first number or frames K1 and less than the second number of frames K2 or not (S13, Fig. 11). If the obtained number of frames K is within the predetermined number, the system determines the series of eye opening and eye closing as a conscious blink (S14, Fig. 11) and ends the blink determination process. On the other hand, if the obtained number of frames K is not within the predetermined number that is below the first number or frames K1 or more than the second number of frames K2, the system determines it as an unconscious blink (S15, Fig. 11) and ends the process.

Figure 12 is a diagram showing the relationship between the number of frames and the blink determination. Even users with weak muscle strength, an unconscious blink (natural blinks) is faster than a conscious blink. The value of the first frame number K1 is set within several to about ten frames in order to eliminate unconscious blinks.

Thus, it is possible to eliminate unconscious blinks that eye opening and closing occur during a few frames from conscious blinks (Fig. 12a). On the other hand, the value of the second frame number K2 is set within about several tens of frames (ex. 50 frames) in order to eliminate the slower blinks than conscious ones. Thus, it is possible to eliminate the continuous closure of eyes that a user with weak muscle strength gets tired to keep his/her eyes on the screen from conscious blinks (Fig. 12b). It is possible to determine conscious blinks adequately by excluding continues closure of eyes and unconscious blinks and regarding a conscious blink when the obtained number of frames is more than the first frame number K1 and lower than the second frame number K2 (Fig. 12c).

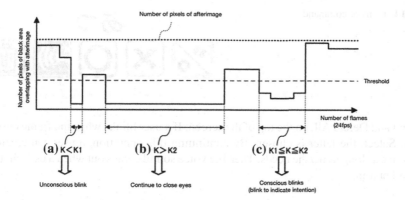

Fig. 12 Relationship between number of frames and blink determination

4 Developing Communication Applications

We developed the blink determination system into new communication applications "Eye Talk" and "Eye Tell" for physically handicapped children.

4.1 Eye Talk

In Eye Talk, a user can select a character in the least number of times. Figure 13 shows the image of the application. A user chooses a consonant in "column" first and a vowel in "row" next from the character table of Japanese of this application.

After choosing a letter, the frame cursor is moving along "Command" items (Fig. 14), which includes Voiced / Semivoiced sound symbols, Delete a letter, Select,

Fig. 13 Character table of Japanese of Eye Talk

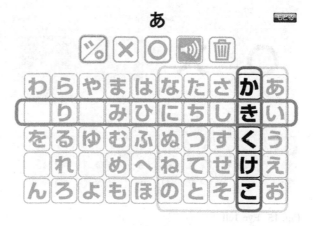

Fig. 14 Items of command

Sound and Delete All, on the top of the screen. If a user blinks when the frame cursor is on Select, the letter is chosen. By continuing this operation, a user can create a sentence as long as he/she needs. Then the voice sound comes out when a user choose Sound button.

4.2 Eye Tell

Eye Tell is a modified communication application to indicate needs simply by "Yes = ○" and "No = ×" by blinks (Fig. 15). A user can make original symbols suitable for the level of handicap. First, a helper makes 2 symbols that are suitable for the situation of a user. (Up to 21 sets.) Then the symbols on right and left sides of the screen turn on and off reciprocally. A user chooses the symbol when it turns on by a blink. The application judges the blink and put the sentence into voice sound.

In Eye Talk and Eye Tell, a user can customize the following settings of the applications.

(1) To show the picture of eyes (On or Off)
(2) Sensitivity of detecting blinks (low 0 to high 100)
(3) Speed of moving the frame cursor on the character table of Jap-anese (slow 0 to fast 100)
(4) Speed of moving the frame cursor on Command items (slow 0 to fast 100)

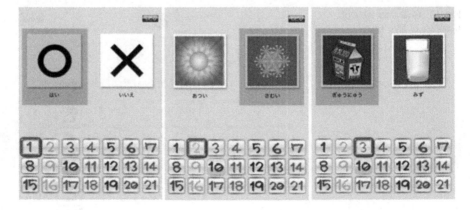

Fig. 15 Eye Tell

The cursor moves $0 = 100$ frames and $100 = 10$ frames. About 24 frames of pictures are processed per a second, so the speed of movement of the cursor is $0 =$ about 4.16 s and $100 =$ about 0.41 s

5 Conclusion

We are on the process of filing a patent for the high quality blink detection system and distributing it free of charge in Japan. For further development of this system, we research and develop the method of determination of gazing directions by analyzing the eye movement. If we can detect the gazing direction accurately on the tablet, burden on a user is reduced. It is applicable to any application by limiting the processing range to achieve higher speed and simplifying the process and high detection accuracy. To determine gazing directions, we use the afterimage method which we have developed in this study. Further, to determine gazing directions, we measure the rate of increase or decrease of the area of the white part of eyes by the afterimage method for processing an image of the eyes. In this way, we can determine gazing directions with high accuracy and less error. By combined the method to determine gazing directions with the blink detection system, a user will be able to get more unfettered communication.

The support applications are living necessities for handicapped people. They should be practical and easy to use. Those applications encourage physically handicapped people to communicate with others and lead the society to support each other regardless of handicap.

References

1. TalkEye: http://www.takei-si.co.jp/product/talkeye.html
2. Let's Chat: http://panasonic.co.jp/phc/products/home/communicationaids.html
3. Viola, P., Jones M.J.: Rapid object detection using a boosted cascade of simple features. In: IEEE CVPR (2001)
4. Lienhart, R., Maydt, J.: An extended set of haar-like features for rapid object detection. In: IEEE ICIP 2002, vol. 1, pp. 900–903 (2002)
5. Tsukada, A.: Automatic detection of eye blinks using spiral labeling. In: Symposium on Sensing via Image Information(SSI 2003), vol. 9, pp. 501–506 (2003)
6. Joblore, G.H., Greenberg, D.: Color Spaces for Computer Graphics, Computer Graphics, vol. 12, pp. 20–25

Author Index

© Springer International Publishing Switzerland 2015
R. Lee (ed.), *Software Engineering Research, Management and Applications*,
Studies in Computational Intelligence 578, DOI 10.1007/978-3-319-11265-7

Printed in the United States
By Bookmasters